Magnetic Sensors: Fundamentals and Applications

Magnetic Sensors: Fundamentals and Applications

Edited by **Casan Anderson**

New York

Published by NY Research Press,
23 West, 55th Street, Suite 816,
New York, NY 10019, USA
www.nyresearchpress.com

Magnetic Sensors: Fundamentals and Applications
Edited by Casan Anderson

International Standard Book Number: 978-1-63238-309-9 (Hardback)

Contents

Preface

The main aim of this book is to educate learners and enhance their research focus by presenting diverse topics covering this vast field. This is an advanced book which compiles significant studies by distinguished experts in the area of analysis. This book addresses successive solutions to the challenges arising in the area of application, along with it; the book provides scope for future developments.

The fundamentals and applications of magnetic sensors are encompassed in this book. It presents an overview on the research conducted in recent years in the field of magnetic sensors. The topics covered in this book vary from underlying theories and properties of magnets to their sensing applications in areas such as biomedicine, microelectromechanical systems, nano-satellites and pedestrian tracking. This book has been compiled for readers who wish to gain a basic understanding of the research domain as well as explore potential areas of applications for magnetic sensors. It also presents novel developments in this field in an efficient and legible manner.

It was a great honour to edit this book, though there were challenges, as it involved a lot of communication and networking between me and the editorial team. However, the end result was this all-inclusive book covering diverse themes in the field.

Finally, it is important to acknowledge the efforts of the contributors for their excellent chapters, through which a wide variety of issues have been addressed. I would also like to thank my colleagues for their valuable feedback during the making of this book.

Editor

Part 1

Theoretical Backgrounds and Principles

Orthogonal Fluxgates

Mattia Butta
Kyushu University
Japan

1. Introduction

Fluxgates are vectorial sensors of magnetic fields, commonly employed for high resolution measurements at low frequencies in applications where the sensor must operate at room temperature.

Fluxgates are usually classified in two categories: parallel and orthogonal fluxgates. In both cases, the working principle is based on a magnetic core periodically saturated in opposite directions by means of an excitation field. The measured field is superimposed to the excitation field and it alters the saturation process.

The basic structure of the orthogonal fluxgate and its difference to parallel fluxgates is illustrated in Fig. 1. The parallel fluxgate (Fig. 1-A) is composed, in its most common form, of a magnetic ring or racetrack core periodically saturated in both directions by the ac magnetic field H_{ex} generated by the excitation coil. The output voltage is obtained with a pick-up coil wound around the core. Even harmonics arise in the output voltage when an external magnetic field H_{dc} is applied in the axial direction. The sensor is called a *parallel* fluxgate because the excitation field H_{ex} and the measured field H_{dc} lay in the same direction. More details about parallel fluxgates can be found in (Ripka, 2001).

Orthogonal fluxgates are based on a similar principle, but they have a different structure, as shown in Fig. 1-B.

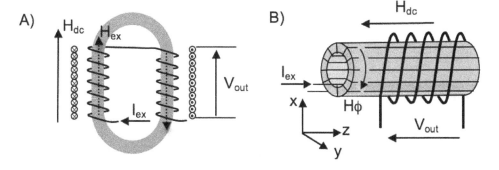

Fig. 1. Structure of parallel (A) and orthogonal (B) fluxgates.

The core is a cylinder of soft magnetic material, with a toroidal excitation coil wound around it. The excitation current flows through the toroidal coil generating excitation Hϕ in the circumferential direction. The core is periodically saturated in the circumferential direction by Hϕ in opposite polarities. Finally, the output voltage is obtained with a pick-up coil as it was in the parallel fluxgate. Once more, when the external field H$_{dc}$ is applied in the axial direction, even harmonics arise in the output voltage. In this case, the sensor is called an *orthogonal* fluxgate because the measured field H$_{dc}$ is orthogonal to the x-y plane where the excitation field Hϕ lays.

Orthogonal fluxgates have been originally proposed in (Alldredge, 1952), both with a cylindrical core and a wire-core. Later, Schonsted proposed an orthogonal fluxgate based on a magnetic wire wound in a helical shape around a conductive wire carrying an excitation current (Schonsted, 1959). Several years later, orthogonal fluxgates appeared again in (Gise & Yarbrough, 1975) where the authors proposed an orthogonal fluxgate with a core obtained by electroplating a Permalloy film on a 6.3 mm diameter glass cylinder after the deposition of a copper substrate. The sensor showed large hysteresis, and it was later improved in (Gise & Yarbrough, 1977) with a core composed of a 3.2 mm diameter copper cylinder and of an electroplated shell on it. An orthogonal fluxgate based on a composite wire, manufactured with a conductive core and electroplated magnetic thin film (about 1 µm thick), was also proposed in (Takeuchi, 1977) after which orthogonal fluxgates were almost forgotten.

From the early years, indeed, parallel fluxgates have been always preferred to orthogonal fluxgates because they usually offer better performances, especially lower noise. Thus, the mainstream of research and development focused on parallel fluxgates.

The development and improvement of techniques for the production of microwires obtained in the last decades (Vázquez et. al, 2011) have made it now possible to manufacture soft magnetic wires with an extremely narrow diameter (50-100 µm) and high permeability. Thanks to this, the principle of the orthogonal fluxgate has been rediscovered. For example, an orthogonal fluxgate sensor based on a glass covered Co-based alloy with a very narrow diameter was proposed in (Antonov et al., 2001) and a similar sensor with a Permalloy/copper wire was used as the core with a 20 µm diameter (Li et al., 2004).

Orthogonal fluxgates based on a microwire gained new popularity mainly due to the rising requests for miniaturized sensors of magnetic fields.

2. Working principle

A detailed explanation of the working mechanism of orthogonal fluxgates is given in (Primdahl, 1970) for the basic tubular structure proposed in (Alldredge, 1958).

Let us consider a tube of soft magnetic material as shown in Fig. 2a, exposed to a sinusoidal excitation field in the circumferential direction Hϕ (generated by a toroidal coil - not shown to simplify the drawing) and to an axial field H$_Z$. The material is assumed to be isotropic with a simplified MH loop shown in Fig. 2b; the magnetization M lies between H$_Z$ and Hϕ in order to satisfy the minimum energy condition.

The axial field is assumed to be much lower than the saturation field H_S, therefore during the part of the period when $H\phi < H_S$ the core is not saturated. Under these conditions, when $H\phi$ increases, then both the angle α and the amplitude of M also increase, while the component of M in the axial direction M_Z does not change because H_Z is constant. However, when $H\phi$ reaches the amplitude where the total field is $H_{tot} = H_S$, then the core gets saturated; if $H\phi$ further increases, then the amplitude of M does not increase anymore and the only effect of $H\phi$ is to rotate M along the circumference which describes the saturation state (Fig. 2c). Under this condition M_Z is not constant anymore but it starts decreasing as the M reaches the saturated state (Fig. 2e). As a result, a variation of the magnetic flux occurs in the axial direction and a voltage is induced in the pick-up coil (Fig. 2d).

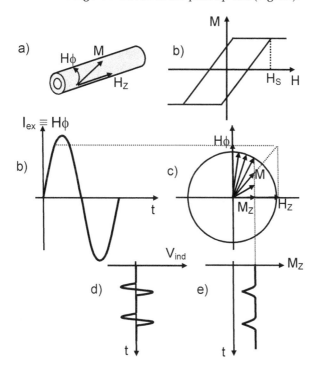

Fig. 2. Working principle of orthogonal fluxgates.

Since the excitation field is sinusoidal the saturation is reached twice per period (i.e. in both the positive and negative directions). This means that the induced voltage will contain even harmonics of the excitation frequency, wherein the second harmonic is generally extracted by means of a lock-in amplifier to obtain the output signal. The amplitude of the induced voltage depends on M_Z, which in turn is determined by H_Z. Finally, the amplitude of the even harmonics gives us a measurement of the axial field H_Z.

If the direction of H_Z is reversed M_Z becomes negative and the phase of the induced voltage is shifted by π rad. This means that the orthogonal fluxgate is able to distinguish between positive and negative fields; usually, the real part of the second harmonic is used as an output signal in order to take into account the phase of the voltage and to obtain an anti-symmetrical function, which allows to discriminate the sign of the field.

2.1 Gating curve

A gating curve is usually measured in order to understand how the flux is gated within the core of a fluxgate. We now consider a real MH loop as in Fig. 3b (without the simplification used in Fig. 2) and we derive the B_Z-$H\phi$ curve that describes the gating occurring in the orthogonal fluxgate core. The amplitude of the peaks in the gating curve is proportional to H_Z since they correspond to M_Z out of saturation. Moreover, the position of the peaks is not constant. For a higher H_Z the saturation is reached for a lower value at $H\phi$, causing the distance between the peaks to decrease (Fig.3).

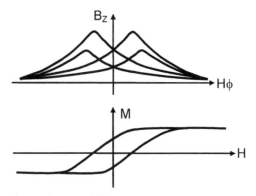

Fig. 3. Gating curves of an orthogonal fluxgate.

The peaks of the gating curve become negative for $H_Z<0$ while no peaks appear when $H_Z=0$. This means that the voltage induced in the pick-up coil is null for no measured field. This becomes extremely important when the sensor is operated in feedback mode with its working point kept around zero. In this case, the output voltage will be always around zero, making it possible to use high gain amplification to increase the signal-to-noise ratio.

2.2 Effect of anisotropy

We must highlight that the model described above applies only if the magnetic core is isotropic or if it has circumferential anisotropy. In case of non-circumferential anisotropy the direction of magnetization M is determined not only by $H\phi$ and H_Z but also by the anisotropy. In this case, the angle θ of M is obtained by minimizing the total energy of M, taking into account the field energy of $H\phi$ and H_Z as well as the anisotropy energy (Jiles, 1991).

Non-circumferential anisotropy can in fact deviate the magnetization from the circumferential plane and give rise to an output voltage even for a zero measured field, significantly changing the gating curves. In such cases, a more detailed model that takes into account the effect of anisotropy should be used (Butta & Ripka, 2008b).

We should also note that in magnetic wires, the anisotropy direction and strength can significantly change according to geometric parameters and manufacturing methods. A detailed characterization of the core's circular and axial magnetic properties is, therefore, always necessary before applying any model to the sensor.

3. Wire-core orthogonal fluxgates

As previously mentioned, the availability of microwires suitable for the fluxgate cores gave new popularity to the orthogonal fluxgate principle.

Fig. 4 shows the structure of an orthogonal fluxgate based on a magnetic wire core. The excitation current I_{ex} is injected to the magnetic wire and generates a circumferential field $H\phi$ while a pick-up coil is wound around the wire as usual.

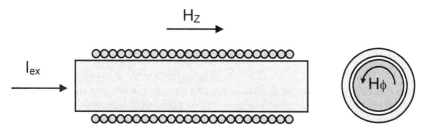

Fig. 4. Orthogonal fluxgate based on a magnetic wire.

In this structure, the excitation coil is not required because the excitation field is generated by the current flowing through the wire. Therefore, the structure of the sensor is extremely simplified and the manufacturing of the sensor becomes easier. Even more importantly, the lack of an excitation coil makes it possible to significantly reduce the dimensions of the sensor. This plays strongly in favor of orthogonal fluxgates, because it makes them suitable for current applications where high miniaturization is required.

Fluxgates based on a microwire became popular also because during the last years the production techniques of magnetic wires have been subject of deep investigation. For example, in (Li et al., 2003) the effect of a magnetic field is shown during the electrodeposition of the NiFe film on a copper wire. By properly tuning the magnetic field's amplitude and direction it is possible to control the anisotropy direction (particularly useful for optimization of sensitivity and offset of the sensors) as well as to improve film uniformity, softness and grain size. Moreover, it has been shown that it is possible to strongly reduce the coercivity of electroplated Permalloy films as well as to increase their permeability by using pulse current instead of dc current for the electroplating process (Li et al., 2006).

Uniformity of the film is improved by a Cu seed layer sputtered on the Cu wire before electroplating because it minimizes the roughness of the surface, helping to reduce the coercivity. The effect of film thickness on the grain size, and finally on the coercivity, has also been studied in (Seet et al., 2006) where it is shown that grain size is lower for larger thickness. However, it is recommended to keep current density constant during the electroplating because if we use a constant current as the thickness increases, the current density decreases, and this is shown to increase the grain size.

3.1 Spatial resolution

Besides the lack of an excitation coil, one of the main advantages of wire-core fluxgates is the diameter of the wire, usually very narrow (several tens of μm). A narrow diameter is advantageous not only for miniaturization, but also for improvement of spatial resolution in magnetic field measurement. Let us consider, for instance, a magnetic field H_Z with constant gradient along the x direction, as shown in Fig. 5. Parallel fluxgates must use either a ring or a racetrack core to reduce the demagnetizing factor and compensate voltage peaks for zero measured fields. Such core has two sensitive sections in the measurement direction (namely A and B in Fig.5, left) which sense different fields H_{ZA} and H_{ZB}. The total field measured by the parallel fluxgate will be the average of H_{ZA} and H_{ZB}.

Parallel fluxgates rarely have a core narrower than 1÷2 cm, limiting the spatial resolution to such level. On the contrary, orthogonal fluxgates have the sensitive cross section of a single wire making it possible to measure the magnetic field H_Z in the single spot, with resolution limited by the diameter. Since typical wires used for orthogonal fluxgates have diameters up to 100 μm, the spatial resolution of orthogonal fluxgates is two orders of magnitude better than conventional parallel fluxgates. To this extent, they were successfully employed for applications such as magnetic imaging. For instance, in (Terashima & Sasada, 2002) a gradiometer based on a wire-core orthogonal flux is presented. The gradiometer is used to measure magnetic fields emerging from a specimen of 3% grain oriented silicon steel, with steps of 50 μm (the diameter of the amorphous wire used as a core is 120 μm). Since the spatial resolution of the sensor is very high it was possible to measure the magnetic field emerging from a single domain, and then graphically represent the domain's topology of the sample.

Parallel fluxgates, based on PCB technology, with an ultra thin core (50 μm) have also been proposed (Kubik et al., 2007). In this case, the spatial resolution is remarkably improved in y direction, but it is still poor in the x direction.

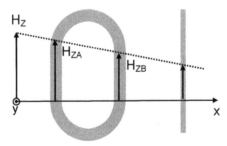

Fig. 5. Spatial resolution in parallel (left) and orthogonal fluxgates (right).

3.2 Excitation field inside the wire

One of the main drawbacks of wire-based orthogonal fluxgates is that the excitation field is not uniform along the distance from the centre of the wire. This comes directly from Ampere's law. Let us consider a magnetic wire with uniform current distribution (i.e. we consider skin effect negligible). The excitation field H_ϕ increases linearly from radius r=0, the centre of the core, to its maximum at the border of the wire (r=R). If we define H_S as the minimum field to saturate the material[1], we observe that the inner part of the wire, for $r<\sigma$, where $H\phi<H_S$ is not fully saturated. On the contrary, when we use a cylindrical core excited by a toroidal coil, then the whole core is equally saturated.

Saturation is a vital requirement for the proper working of a fluxgate, wherein only the outer saturated shell will contribute to fluxgate mode whereas the inner unsaturated part of the core will not act as a fluxgate. Most important, having the central part of the core unsaturated causes hysteresis in the output characteristic of the fluxgate. Indeed, if we apply an axial magnetic field to the wire this will magnetize the central part of the core in its direction. Since that part of the core is not saturated, the magnetization cannot be restored by the excitation field through saturation in the circumferential direction. The centre of the core will then naturally follow its hysteresis loop.

To this extent, it is very important to achieve the full saturation of the core to avoid the hysteretic behaviour of the sensor. Unfortunately, it is impossible to saturate the wire in its entire cross-section, since this would require an infinite current. Instead, we will always have an inner portion of the wire unsaturated.

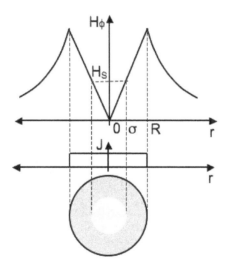

Fig. 6. Magnetic wire with uniform current distribution. The magnetic field increases linearly within the wire and only the outer shell where $H\phi>H_S$ is saturated.

Amorphous wires are often used as cores for orthogonal fluxgates. In this case, the wire has an inner cylinder with magnetization in the axial direction and a shell with radial or circumferential magnetization (Fig. 7) in case of positive or negative magnetostriction respectively (Vázquez & Hernando, 1996).

---

[1]The saturation field is clearly not a brick wall border. The amount of saturated material asymptotically increases when the magnetic field grows. Therefore, we cannot define a clear border between the saturated and unsaturated state. However, we can define a condition when the core can be considered saturated from a practical point of view. That occurs when any increment of the magnetic field does not cause any significant change in the working mechanism of the fluxgate.

In this case the central part of the core will never contribute to the fluxgate effect, which will be given only by the outer shell. The inner part of the core usually shows a bistable behaviour, which means that its magnetization will switch direction upon the application of an axial field larger than the critical field. A fluxgate base on such wires will be affected by the perming effect (i.e. shift of the sensor's output characteristic after the application of a large magnetic field) due to the switching of the magnetization in the central part of the wire.

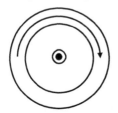 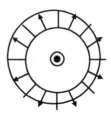

Fig. 7. Cross-section of a magnetic wire with bamboo structure, in case of negative (left) and positive (right) magnetostriction.

3.3 Composite wires

Composite wires have been proposed to solve problems given by the unsaturated inner section of the wire (Ripka et al., 2005; Jie et al., 2006). The main idea involving composite wires is to have wires with non-magnetic cores surrounded by a soft magnetic shell. In this way, we avoid problems such as the hysteresis of the sensor's characteristic and the perming effect, which typically arise if the wire is not fully saturated.

Considering a core composed of a 20 μm diameter copper wire surrounded by a 2.5 μm thick Permalloy layer, the perming error (i.e. shift of offset after 10 mT shock field) is only 1 μT, for an excitation current as low as 20 mA. Moreover, it is shown that the perming error decreases for a higher excitation current, as typically found for bulk core fluxgates, because the core is more deeply saturated.

The most frequently used technique to produce composite wires consists of the electroplating of a magnetic alloy, for example $Ni_{80}Fe_{20}$ (Permalloy), on a copper microwire. The resistivity of copper (~17 nΩ·m) is lower than the resistivity of many magnetic alloys (for instance the resistivity of Permalloy is ~200 nΩ·m). For a typical wire composed of a 50 μm diameter core and surrounded by a 5 μm Permalloy shell, only 3.6% of the total current flows through the magnetic shell. If we operate the sensor with an excitation current low enough to make skin effect negligible, we can assume that the whole excitation current will flow through the copper core. Such simplified configuration is shown in Fig. 8 where the current density J is uniform within the copper core and zero in the magnetic shell. The circumferential magnetic field generated by the excitation current linearly rises until within the copper core ($r=R_c$) and then it decreases as $1/r$ for $r>R_c$ (i.e. on the magnetic shell). In this case, the outer part of the magnetic layer is excited by a lower field, namely H_m. As far as the excitation current is high enough to make $H_m>H_S$ we can consider the wire completely saturated.

In this kind of structure, a larger magnetic layer requires a larger excitation current in order to avoid that the outer portion of the magnetic shell becomes unsaturated. Therefore, we must carefully weigh the advantages of lager sensitivity given by a thicker magnetic shell against the disadvantages caused by an increment of current required for the saturation.

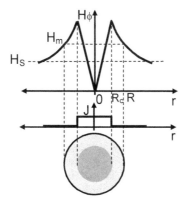

Fig. 8. Composite wire with copper core and magnetic shell. The current flows entirely through the copper core so that the magnetic shell is fully saturated.

Skin effect, however, is not always negligible, especially when the sensor is operated at a high frequency in order to increase the sensitivity. In this case, the excitation current drains from the copper core to the magnetic shell, reducing the magnetic field in the magnetic shell. Depending on the actual current distribution, the magnetic field can strongly change. Numerical simulation is usually employed in order to predict the current distribution within composite wires (Sinnecker et al., 2002). The penetration depth strongly depends on the conductivity of both the conductive core and the magnetic shell as well as on the permeability of the latter. Therefore, a general value for a limit frequency to avoid draining the current to the magnetic shell cannot be given. Numerical simulation is suggested to predict current distribution within the wire.

Finally, designers of orthogonal fluxgates should carefully choose their operating frequency. On the one hand, a higher frequency increases the sensitivity, which contributes to the reduction of noise, whereas on the other hand, a higher frequency can cause parts of the wire not to be completely saturated, incrementing the noise (besides the hysteresis and perming effect). The excitation frequency should be chosen as a compromise between these two opposite effects.

3.4 Glass insulation

A more complex structure has been proposed by (Butta et al., 2009a) to overcome the problem of the current draining to the magnetic shell due to the skin effect. This is carried out by putting a glass layer between the copper core and the magnetic shell. The glass layer provides electrical insulation, helping thus to keep the excitation current flowing entirely within the copper core, regardless of the operating frequency. Even if the skin effect should occur in the copper core, given Ampere's law, this does not affect the magnetic field generated from the copper's diameter.

In order to manufacture a composite structure with glass insulation between the copper and the magnetic shell, glass coated copper wires are used as a base. Following this procedure, a small nm thick gold layer is applied on the glass coating by means of sputtering. Finally, the electroplating of magnetic alloy is performed on the gold seed layer.

By using such structure the saturation current can be strongly reduced. In (Butta et al., 2009a), the saturation current is reduced by a factor of 3.

4. Micro orthogonal fluxgate

As already mentioned, the lack of an excitation coil is one of the main advantages of orthogonal fluxgates, because it strongly simplifies its structure, making high miniaturization possible.

The first attempt made in order to reduce the dimensions of an orthogonal fluxgate was carried out in (Zorlu et al., 2005) where a sensor is based on a wire composed of Au core (20 μm diameter) covered by a 10 μm thick FeNi electroplated layer. The total diameter of the wire is therefore 40 μm, and the length varies from 0.5 to 4 mm. The output voltage is picked-up by means of two planar coils fabricated on a Pyrex substrate by means of sputtering, photolithography and patterning.

The response of the sensor has a large linear range for excitation current, which can be as low as 50 mA (at 100 kHz), showing that the wire is saturated for such low current. If the current is further increased to 100 mA the linear range reaches ±250 μT, and sensitivity reaches 4.3 V/T. A higher current than the minimum current necessary to saturate the core is also useful against the perming effect. While perming shift after ± 50 μT shock field is 16 μT for 50 mA excitation current, it drops down to 2 μT for 100 mA excitation current.

Orthogonal fluxgates based on a microwire, however, can hardly be manufactured at lower dimensions. The microfabrication of the sensor becomes more suitable for micro sensors, especially for mass production. In (Zorlu et al., 2006) a microfabricated orthogonal fluxgate is presented wherein the core is manufactured in three steps. First, a Permalloy bottom layer is electroplated on the Cr/Cu seed layer previously applied on the substrate, then the central copper core is electroplated in the middle and finally Permalloy is electroplated on the three open sides of the copper creating a closed loop of Permalloy around the copper. The resulting structure is composed of a rectangular shape core (8 μm x 2 μm) and a copper nucleus surrounded by a 4 μm Permalloy layer (the total dimensions of the structure is 16 μm x 10 μm). The length of the core is 1 mm. The dimension of the core was finely adjusted thanks to the high precision of photolithography.

Also in this case, the flux is picked-up using two planar coils formed in the substrate under the core (2 x 60 turns). The sensor has a large linear range (±200 μT) but rather low sensitivity, around 0.51 V/T for a 100 mA excitation current at 100 kHz. Thus, the resulting noise is higher than typical orthogonal fluxgates (95 nT/√Hz at 1 Hz). One of the problems of such configuration is that the planar coils cannot properly pick-up the flux as a concentric coil. Clearly, further investigation is necessary to understand whether a different configuration of the coil can significantly increase the sensitivity and then reduce the noise.

5. Multi-wire core

One of the main drawbacks of orthogonal fluxgates based on magnetic wires is low sensitivity, mainly due to cross-sectional areas lower than traditional parallel fluxgates or orthogonal fluxgates based on bulk tubular cores.

In order to increase the sensitivity, multi-core sensors have been proposed wherein the core is composed of multiple magnetic wires closely packed, each of them excited by a current with equal amplitude and frequency. The wires are also not electrically in contact along their length. In case of amorphous wires, a thin glass coating (typically 2 μm) provides insulation between them. For composite Cu/Py wires a small nm layer of epoxy is added to the surface of the wire to assure insulation.

In (Li et al., 2006a) the sensitivity of a multi-wire core fluxgate with tuned output was measured for cores with a different number of wires and it was found to increase exponentially; for instance, a 16 wire core has sensitivity 65 times higher than the sensitivity of a single wire. Later on it was demonstrated (Li et al., 2006b) that such growth of sensitivity was not simply caused by the increase of ferromagnetic material composing the core. Let us consider a sensor having a single wire core and a sensor based on a two-wire core whose total cross-sectional area is comparable to the area of a single wire. In such a case, the sensitivity is higher for the two-wire core despite the cross sectional area being similar to the single sire core. It is shown that the increment of the sensitivity becomes linear if the wires are kept far enough (5 time the diameter). This suggests the cause of the exponential increment of sensitivity for multi-wire cores is the magnetic interaction between the wires.

An increment of sensitivity is, however, useless if the noise also increases. Further investigation (Jie et al., 2009) has proven that orthogonal fluxgates with a multi-wire core do not only have higher sensitivity but also lower noise. It is interesting to note that the noise is lowest for configurations where the wires are arranged in the most compact way, because the mutual interaction between the wires is stronger the closer they are. Therefore, multi-wire cores are convenient both in terms of sensitivity and in terms of noise.

Later (Ripka et al., 2009) suggested that the exponential increment of the sensitivity to the number of wires is due to the improvement of the quality factor of the tuning circuit. This was then confirmed in (Ripka et al., 2010) where the anomalous increase of sensitivity is explained to be due to changes of parametric amplification caused by changes in the quality factor of the tuning circuit.

The total cross-sectional area is clearly higher for multi-core fluxgates and, therefore, the spatial resolution is worse than the single wire core. However, we should consider that the sensitivity increases exponentially, meaning that the sensitivity per unit of area is higher in multi-wire cores. In any case, if we consider a 16 wire core, the spatial resolution decreases by a factor of ~4, depending on the geometry of the configuration. This is still one order of magnitude better than sensors based on bulk cores.

Another advantage of a multi-wire core is the mutual compensation of spurious voltages if wires are connected in an anti-serial configuration. As an example, two-wire core has 0.34 nT/√Hz noise at 1 Hz.

Finally, we must be careful about the interaction that may occur between the wires if closely packed. This might cause hysteresis in the response of the sensor for low field measurements (Ripka et al., 2010).

6. Fundamental mode

Orthogonal fluxgates have been ignored in the past because they have higher noise than parallel fluxgates. This, in fact, moved the mainstream of research to focus on parallel fluxgates, since noise is one of the most important parameters for high precision magnetometers (other parameters such as linearity or sensitivity can be compensated to a large extent by proper design of electronics or sensors). Despite the fact that orthogonal fluxgates have recently gained new popularity due to their high spatial resolution and simple structure, their noise is still an issue for these kinds of sensors. Micro fluxgates are reported to have noise around units of nT/√Hz at 1 Hz, while wire core orthogonal fluxgates typically have 100÷400 pT/√Hz noise at 1 Hz. Without substantial reduction of noise, orthogonal fluxgates cannot be considered competitive to parallel fluxgates.

An important step forward in the field of noise reduction in orthogonal fluxgates was made by Sasada, who proposed to operate the sensor in fundamental mode rather than in second harmonic mode (Sasada, 2002a).

6.1 Working mechanism

The structure of the sensor is identical to the wire-core orthogonal fluxgate; however a dc bias is added to the excitation current. The output voltage induced in the pick-up coil in this case will be at a fundamental frequency.

In order to understand the working mechanism underlying fundamental orthogonal fluxgates we can refer to Fig. 9. Since a dc bias is added to the excitation current, the resulting excitation field in the circumferential direction turns out to be as follows:

$$H\phi = H_{dc} + H_{ac} \sin (2 \cdot \pi \cdot f \cdot t)$$

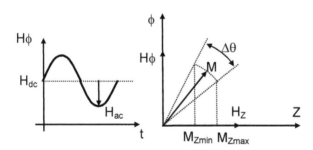

Fig. 9. Schematic diagram of the working mechanism of orthogonal fluxgate operated in fundamental mode.

The dc bias must be large enough to make the excitation field unipolar. As a result, magnetization won't reverse its polarity, as for a symmetrical bipolar excitation current with no dc bias. The magnetization M will oscillate between $\pm\Delta\theta/2$ in order to always satisfy the minimum energy condition, taking into account the field energy of $H\phi$ and H_Z as well as anisotropy energy. In traditional fluxgates without the dc bias, the magnetization is reversed from positive to negative saturation and vice versa each period, thus the output voltage contains mainly a second harmonic. On the contrary, in the fundamental mode orthogonal fluxgate the dc bias does not allow the magnetization to reverse polarity but only to oscillate with the same frequency f of $H\phi$. Therefore, the output voltage induced in the pick-up coil by time varying M_Z (component of M in Z direction) will be sinusoidal at a fundamental frequency.

At this point, we should point out that this sensor must be, after due consideration, classified as a fluxgate sensor, despite some similarities with other sensors. The magnetic flux within the core is indeed still gated; the sensor works at best, returning a linear and a bipolar response when the excitation field is large enough to deeply saturate the core, as typically found in fluxgates. The only difference between traditional fluxgates without dc bias and fundamental mode orthogonal fluxgates is that the flux is gated only in one polarity rather than in both polarities.

6.2 Offset

So far we have not discussed the effect of anisotropy on the output voltage. The anisotropy contributes to determine the position of magnetization. For instance, if $H_Z=0$ the resulting M_Z is null only if $\alpha=\pi/2$ (i.e. if anisotropy is circumferential). Contrarily, if the anisotropy is non circumferential (i.e. $\alpha<\pi/2$, as in Fig. 10) then M_Z will be non-zero even for $H_Z=0$ and M will lie between $H\phi$ and K_u ($\alpha<\theta<\pi/2$). As a result, the output voltage due to time variation of M_Z will be non-zero despite $H_Z=0$. This means that the sensor's response will show an offset anytime the anisotropy is not circumferential.

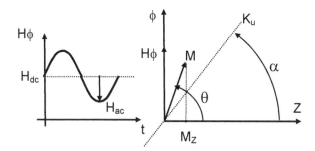

Fig. 10. Non-circumferential anisotropy in a magnetic wire used as core for fundamental mode orthogonal fluxgates.

Unfortunately, non-circumferential components of anisotropy are typically found both in amorphous wires and composite Cu/Py wires. The output offset is therefore always expected in fundamental mode orthogonal fluxgates. In order to suppress the offset, a technique is proposed in (Sasada, 2002b). Sasada's method is based on the fact that the sign of the characteristic is reversed if the dc bias becomes negative, while the offset is unchanged. For $H_Z=0$ the magnetization M will oscillate around θ_0' for positive dc bias and around θ_0'' for negative dc bias (Fig. 11). The projection of M on the Z axis will be identical because $\theta_0''=\theta_0'+\pi$ and H_{ac} makes M rotate in the opposite direction according to the bias sign.

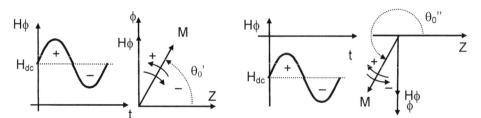

Fig. 11. Diagram of fundamental mode orthogonal fluxgates with positive and negative dc bias. The signal sensitivity is inverted changing the sign of dc bias but the offset is unchanged.

In order to suppress the offset we can periodically invert the dc bias and subtract the signals obtained with the positive and the negative bias. Since the sensitivity is reversed, by subtracting the characteristics we sum up the signals whereas the offset is cancelled given the fact that its sign is unchanged for both the positive and the negative bias.

The bias can be switched at a frequency much lower than the excitation frequency. For example, (Sasada, 2002b) suggests to invert the sign every 25 periods of excitation current. In this way we can reduce the effect of sudden transition from a saturation state to an opposite saturation state which could negatively affect the output noise of the sensor. To avoid the effect of bias switching on the noise we can exclude the period right before and after the transition. This can be easily done digitally (Weiss et al., 2010) or analogously using a fast solid state switch before the final low pass filter (Kubik et al., 2007).

It must be noted that all the proposed techniques require significant modification of the electronics both on the excitation side as well as on the signal conditioning circuit. While this slight complication in the electronics can be bearable for many magnetometers, it could be a non-negligible problem for applications such as portable devices.

6.3 Noise

Orthogonal fluxgates in fundamental mode became very popular thanks to the fact that they have less noise than traditional orthogonal fluxgates. This is due to their operative mode, rather than the sensor itself. In (Paperno, 2004) it is demonstrated how the very same fluxgate (120 µm diameter Co-based amorphous wire surrounded by 400 turn pick-up coil) has $1 \text{ nT}/\sqrt{\text{Hz}}$ noise at 1 Hz if operated in the second harmonic mode whereas the noise is reduced to $20 \text{ pT}/\sqrt{\text{Hz}}$ when the fundamental mode is used. In this case, the fundamental mode contributes to reduce the noise by a factor of 50, obtained using the same sensor.

A similar result was obtained in (Paperno et al., 2008) for a fluxgate based on a tubular core manufactured with a 5 cm wide amorphous ribbon wrapped with 8 mm of outer diameter. In this case, both the excitation and pick-up coils are added to the core. When this sensor is operated in a fundamental mode, the noise results as being $10 \text{ pT}/\sqrt{\text{Hz}}$ at 1 Hz, or 30 times lower than the value obtained in the second harmonic mode.

Therefore, noise reduction given by the fundamental mode can be generalized as it applies to all kinds of orthogonal fluxgates, based on the wire core as well as on bulk tubular core.

This can be easily seen when analyzing the source of the noise. Typically, the noise of fluxgate sensors originates in the magnetic core. The reversal of magnetization from positive to negative saturation (and vice versa) involves domain wall movement, which is the origin of the Barkhausen noise. Since a pick-up coil detects time-variation of flux within the core, the Barkhausen noise will cause noise in the output voltage of the pick-up coil. Therefore, designers of fluxgates have chosen materials for the core, which are not only very easy to saturate but also present very smooth transitions between opposite saturation states.

This source of noise is dramatically reduced when a dc bias is added to the excitation current. If the bias is large enough to keep the core saturated for the whole period of the ac current I_{ac}, then the magnetization is only rotated by I_{ac} (Fig. 9) and no domain wall movement occurs.

Sensitivity, however, should also be considered when calculating the output noise in magnetic units. A higher dc bias I_{dc} can significantly reduce sensitivity, because it increases the angle θ of magnetization M resulting in a lower projection of M on the longitudinal axis (i.e. the magnetic flux in the longitudinal direction is sensed by the pick-up coil). On the contrary, the sensitivity monotonically increases with the ac excitation current I_{ac} (Butta et al., 2011) and therefore an increment of I_{ac} can be useful to reduce the total noise even if a larger I_{ac} could bring the core out of saturation.

The lowest noise of an orthogonal fluxgate in fundamental mode is then obtained selecting a pair of parameters I_{ac} and I_{dc} such that the sensitivity is large enough to minimize the noise but with the minimum value of the total current not too low, so as to avoid significant domain wall movement in the core. The optimum condition for noise reduction is obtained right before minor loops appear in the circumferential BH loop (Butta et al., 2011). Noise as low as 7 pT/$\sqrt{Hz}$ at 1 Hz was obtained by optimizing excitation parameters, using the magnetometer structure proposed in (Sasada & Kashima, 2009).

The noise can be further reduced to 5 pT/$\sqrt{Hz}$ at 1 Hz by using three-wire cores instead of a single wire, in order to increase the sensitivity.

7. Coil-less fluxgates

As previously mentioned, orthogonal fluxgates based on microwires gained popularity due to the absence of the excitation coil, which help to simplify the manufacturing process. To this extent, the wire-core needs only a pick-up coil, which can be easily wound around it with an automatic procedure. However, the presence of a coil, even if it is simply a pick-up coil, can make the sensor unsuitable for applications where high miniaturization is required. A possible solution to this problem is to use planar coils manufactured on a substrate under the fluxgate core as in (Zorlu et al., 2006) although this solution has a more complicated structure, which needs an extra step in the manufacturing process. It would be better to have a fluxgate without any pick-up coil at all. This can be achieved with coil-less fluxgates (Butta et al., 2008a).

7.1 Structure of the sensor

In a coil-less fluxgate, torsion is applied to a composite microwire with a copper core covered by a ferromagnetic layer, while an ac excitation current flows through the wire (Fig.

12). If the excitation current is large enough to saturate the magnetic layer in both polarities and a magnetic field is applied in the axial direction, then even harmonics will arise in the voltage across the terminations of the wire. It was found that a second harmonic is proportional to the magnetic field applied in the axial direction; therefore this structure can be used as a magnetic sensor. Since the output voltage is obtained directly at the terminations of the wire no pick-up coil is required.

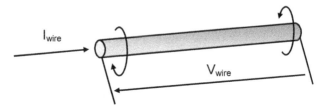

Fig. 12. Coil-less orthogonal fluxgate. The magnetic wire is twisted and the output is obtained at the wire's terminations.

It should be noted that this sensor must be classified, after due consideration, as an orthogonal fluxgate, even if the structure could recall that of magneto impedance (MI) sensors. Indeed, the sensor returns an output signal with linear characteristic only if full saturation of the wires is achieved in both polarities; if saturation is lost the signal vanishes. Moreover, the operative frequency for a coil-less fluxgate is around 10 kHz, whereas MI sensors are operated at MHz range. This means that the physical phenomena occurring within the wire are substantially different. In other words, MI sensors are mainly based on variation of skin effect in the magnetic wire due to a change of permeability caused by the external field (Knobel et al., 2003) whereas in coil-less fluxgates the external field causes linear shifting of a circumferential BH loop, giving rise to even harmonics. The difference between sensors becomes evident when considering their output characteristics. Coil-less fluxgates, have a second harmonic, which linearly depends on the external field with anti-symmetrical characteristic. This allows one to discriminate between positive and negative fields. MI sensors have, on the other hand, impedance, which shows a non-linear symmetric characteristic. In order to be used in a magnetometer, MI sensors must be biased with a dc field (Malatek et al., 2005), so that the working point will move in the descendent branch of the characteristic (the output, however, will only be approximate to a linear function).

7.2 Working mechanism

In (Butta et al., 2008a) it is shown how the sensitivity of a coil-less fluxgate depends on the twisting angle applied to the magnetic wire and how the sensitivity becomes negative if the wire was twisted in the opposite direction. No output signal was instead recorded for no twisting applied to the wire. Therefore, it was assumed that the working mechanism of the coil-less fluxgate took place due to helical anisotropy induced into the magnetic wire by mechanical twisting. This was later confirmed by observing coil-less fluxgate effect also on magnetic wires manufactured with built-in helical anisotropy. In (Butta et al., 2010b) a Permalloy layer is electroplated under the effect of a helical field, obtained as a combination of a longitudinal field imposed with a Helmholtz coil and a circular field generated by a dc current flowing in the wire. In (Atalay et al, 2011; Butta et al., 2010c; Kraus et. al, 2010) helical anisotropy is induced in the wire electroplating the Permalloy under torsion and releasing it at the end of the manufacturing process. The back-stress after such release is responsible for helical anisotropy.

In (Butta & Ripka, 2009b) a model for the working mechanism of the coil-less fluxgate is proposed, based on the effect of helical anisotropy on the magnetization of the magnetic wire, during the saturation process determined by the excitation current.Fig. 13 shows the circumferential BH loop (Ripka et al., 2008) of the magnetic wire with +80 µT, - 80 µT, and 0 µT of the external field applied to the axial direction.

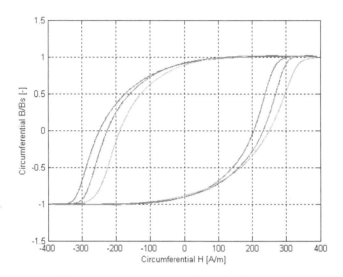

Fig. 13. Circumferential BH of a magnetic wire with applied torsion for 0 µT and ±80 µT field applied in the axial direction. The loop is shifted by the external field.

The circumferential flux is obtained by the integration of the inductive part of the voltage across the wire's terminations V_{wire}. In turn, the inductive component of V_{wire} is obtained subtracting the resistive part of the voltage calculated as $R_{wire} \cdot I_{wire}$. The voltage measured on the terminations V_{wire} will then be the derivative of the circumferential flux; when the magnetization is reversed from positive to negative saturation and vice versa, the voltage peaks appear in V_{wire} in addition to the resistive voltage drop.

Let us consider a microwire with helical anisotropy as shown in Fig. 14, where γ is the angle axis of easy magnetization in regards to the axial direction of the wire Z. As observed in cases of traditional fluxgates, the magnetization is rotated by the excitation field $H\phi$, which periodically saturates the wire in the opposite direction. However, the mechanism is now rotated by an angle γ. Therefore, the field responsible for the rotation of M is now the component of $H\phi$ perpendicular to the easy axis of magnetization, namely $H\phi\perp$. The dc axial field also has a component on the perpendicular axis, $H_Z\perp$, which acts as a dc offset to the ac $H\phi\perp$. This implies that the periodical process of saturation caused by the excitation field is shifted by the axial field through its component $H_Z\perp$. If we observe the circumferential BH loop using $H\phi$ as a reference, then we observe a shift of the loop under the effect of the axial field as shown in Fig. 13. The sensitivity of the sensor increases together with the increasing anisotropy angle γ because the higher is γ the larger is $H_z\perp$.

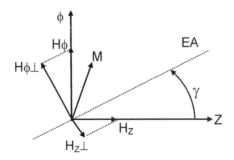

Fig. 14. Working mechanism of coil-less fluxgates.

7.3 Sensitivity

The sensitivity of coil-less fluxgates strongly depends on the amplitude of the excitation current. However, while the sensitivity of traditional fluxgates increases if we use a bigger current, the sensitivity of coil-less fluxgates decreases. This means that the higher the excitation current is, the lower the sensitivity will result (Fig. 15). This can be clearly explained by considering the model of the sensor. By increasing the excitation current, the field energy associated to the circular magnetic field will also increase, causing the magnetization M to be tied more strongly to the excitation field in a circular direction, while the effect of anisotropy energy on the total energy of M will become progressively negligible.

By observing Fig. 15 one might think that the best working condition for coil-less fluxgates is obtained with an excitation current about 42÷43 mA, where the sensitivity is at its maximum. However, the excitation current must be high enough to fully saturate the wire, in order to lower the noise as well as to assure a wider linear range. Since an external field shifts the circumferential BH loop of the magnetic wire (Fig.13), the sensor will keep working regularly as long as the measured field is not too large to move one end of the BH loop out of saturation. If that were to happen, the linearity of the sensor would be lost. Therefore, it is recommended to keep the sensor working at a higher excitation current than the minimum current required to achieve saturation, although still not high enough to avoid significant loss of sensitivity.

Compared to traditional fluxgates a coil-less fluxgate has generally lower sensitivity. This is due to the fact that we pick up the circumferential flux with a virtual one-turn coil. While fluxgates with a pick-up coil can simply multiply the sensitivity by using a large number of turns, this is not possible for coil-less fluxgates.

Typical sensitivity for coil-less fluxgates based on a composite Cu-Permalloy wire is about 10 V/T. This value is significantly higher if a Co-based wire is used. In (Atalay et al., 2010) it is reported that a coil-less fluxgate obtained with a Co rich amorphous wire after 15 minutes joule annealing, which reaches sensitivity at about 400 V/T at 30 kHz. In (Atalay et. al, 2011) a coil-less fluxgate based on a composite copper wire with $Co_{19}Ni_{49.6}Fe_{31.4}$ electroplated shell is proposed. The sensitivity in this case is about 120 V/T at 20 kHz. Further research on different materials will show if even higher sensitivity will be achievable with other alloys.

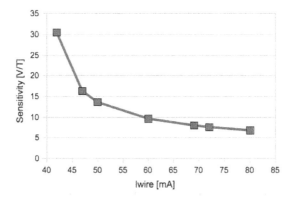

Fig. 15. Output characteristic of coil-less fluxgates for different amplitudes of excitation current I_{wire}. The higher I_{wire} becomes, the lower the sensitivity of the sensor will be.

Sensitivity can be also increased with higher angles of helical anisotropy but we should keep in mind that saturation current also increases, and this will require a higher excitation current.

A drawback of coil-less fluxgates is that low sensitivity cannot be increased by using high gain amplification because the output voltage of the sensor includes large spurious voltage. This component of the voltage does not include a signal but contributes to enlarge its peak value, limiting the maximum amplification. The resistive part of the spurious voltage, due to the voltage drop on the wire's resistance can be easily removed by a classical resistive bridge. However, the inductive component of the voltage, given by the transition of the magnetization from one saturated state to the opposite, will be always present in the output. As previously explained, these peaks will be shifted by the external field to opposite directions, but they will continue to be present in the output. A technique proposed by (Butta et al., 2010a) is presented to remove the inductive peaks and obtain an output voltage that is null for no applied field and whose amplitude increases proportionally to it. The method is based on a double bridge with two sensing elements fed by current in opposite directions. In the output voltage, the positive peaks of the first wire will be compensated by the negative peaks of the second wire and vice versa. The sensitivity of the two wires must clearly point to opposite directions so that the sum of the voltage obtained with the opposite current will be the sum of the two signals rather than their difference.

7.4 Linearity

A common technique used to improve linearity of magnetic sensors is to operate them in a closed loop mode, by generating a compensation field, which nullifies the measured field (Ripka, 2001). The pick-up coil is usually used for this purpose, because the compensating field must be generated at a low frequency, several orders of magnitude lower than the excitation frequency. Using the feedback mode, the working point of the sensor will always be around zero magnetic field and the output characteristic will be determined by the linear characteristic of the coil.

This method, however, cannot be used for coil-less fluxgates, since it has not a pick-up coil available for the generation of a compensating field (and if we add a compensation coil the sensor would not be coil-less anymore).

Therefore, the linearity of the coil-less fluxgate is an extremely important parameter, because the sensor will be used in an open loop mode. Fortunately, the coil-less fluxgate has a large linear range. In (Butta et al., 2010c) it is shown that a coil-less fluxgate with ±0.5% of full-scale non-linearity error in a ±50 μT measurement range. The non-linearity error is reduced to ±0.2% of full scale if we consider a ± 40 μT range. These values are comparable to the non-linearity of non-compensated parallel fluxgates (Kubik et al., 2009; Janosek & Ripka, 2009).

The high linearity of coil-less fluxgates comes from the working mechanism of the sensor, which is simply based on linear shifts of the circumferential BH loop. Non-linearity might be due to the non-uniformity of the helical anisotropy angle along its length. Further improvements of the manufacturing process can help make the anisotropy more uniform and improve the linearity of the sensor.

7.5 Noise

The noise of a coil-less fluxgate is rather high. For instance, in (Butta et al., 2010c) a coil-less fluxgate is presented which shows 3 nT/√Hz at 1Hz noise. This is much higher than the noise of other orthogonal fluxgates, operated in a fundamental mode, mainly because of low sensitivity. The noise of coil-less fluxgates manufactured with Co-base magnetic wires, which have larger sensitivity, has still not been reported. It can be expected that further improvements of the sensitivity of coil-less fluxgates will contribute to decrease the noise.

8. Comparison

It is important to understand both the advantages and disadvantages of orthogonal fluxgates when we have to select a magnetic sensor for a specific application. Depending on the particular requirements of the measurement system, the best solution can be a parallel or an orthogonal fluxgate. Here we give a list of both advantages and disadvantages of orthogonal fluxgates in order to help the user in choosing the best sensor for his/her purposes.

Advantages of orthogonal fluxgates

- high spatial resolution, limited by the wire diameter (usually around 100 μm);
- lack of excitation coil, which implies a smaller structure;
- easy to manufacture;
- low excitation current (many wires require a few tens of mA to be saturated, whereas parallel fluxgate cores are often saturated with several hundreds of mA).

Disadvantages of orthogonal fluxgates

- higher noise than parallel fluxgates;
- lower sensitivity due to small cross-sectional areas of the wire-core (this can be increased by using a multi-wire core to the expense of the spatial resolution);

- the excitation current flowing directly to the wire-core generates power dissipation within the wire; this can increase the temperature of the wire causing dilatation and finally mechanical stress, which is a typical source of noise.

The following table summarizes several orthogonal fluxgates reported in the literature with their features and obtained performance. The proper choice for structure and operative parameters of orthogonal fluxgates can be made based on the application requirements and available performances summarized here.

	Sasada. 2009	Zorlu, 2007	Paperno, 2004	Fan, 2006	Li, 2006	Goleman, 2007
Principle	Fundamental mode	Second harmonic	Fundamental mode	2nd harmonic (tuned)	2nd harmonic (tuned)	Fundamental mode
Configuration	U-shaped amorphous wire	Planar Cu/Permalloy structure	Amorphous wire	Cu/Permalloy Wire	16 glass coated amorphous wires	U-shaped amorphous wire
Length	40 mm (20 mm sensitive length)	1 mm	20 mm	9 mm	18 mm	28 mm
Diameter	120 μm	16 μm x 10 μm (squared)	120 μm	20 μm	16 μm	125 μm
N. of turns pick-up coil	2 coils x1000 turns	2 planar coils x60 turns	400	1000	1000	250
Excitation Current	8mA ac + 47 mA dc	100 mA peak sinusoidal	40 mA ac + 40mA dc	10 mA rms sinusoidal	6 mA rms sinusoidal (each wire	4 mA ac + 20 mA dc
Frequency excitation	118 kHz	100 kHz	40 kHz	500 kHz	188 kHz	100 kHz
Sensitivity	350,000 V/T (gain 47)	0.51 V/T		20,000 V/T	850,000 V/T	1,600 V/T
Offset	-0.33V				48.2 mV	
Linear range	±25 μT	±100 μT				
Noise PSD @ 1 Hz	10 pT/√Hz	95 nT/√Hz				0.11 nT√Hz at 10 Hz
Resolution		215 nT	100 pT			
Power consumption		8.1 mW		100 mW		

Table 1. Comparison of several types of orthogonal fluxgates

9. Future development

During this last decade, the research has been focused mainly on issues regarding orthogonal fluxgates, like noise reduction, increment of sensitivity, and simplification of the sensors' configuration and development of wires with new structures.

These efforts strongly improved the performances of orthogonal fluxgates, making this sensor competitive in the field of magnetic measurement at room temperature.

However, even if sensors like orthogonal fluxgates in a fundamental mode already achieved noise levels similar to cheap parallel fluxgates, other issues have to be faced.

Currently, we still do not have extensive information about the long-term offset stability of orthogonal fluxgates as well as the temperature dependence of both offset and sensitivity, which are critical points for many magnetometers.

Another important field, which has to be investigated, is the dependence of the orthogonal fluxgate's performance on the geometrical dimensions of the core. So far, different structures have been proposed, but a comprehensive study that explains the effect of different core sizes on sensitivity and noise has yet to be reported. In particular, the effects of the demagnetization factor have not been properly investigated, mainly due to the fact that the excitation field is applied to a circumferential direction facing a toroidal shape, which is not affected by the demagnetizing effect. Nevertheless, a measured field is applied in the axial direction over a finite length specimen so that the internal field distribution will be affected by the demagnetizing effect. This applies especially to multi-core orthogonal fluxgates. Indeed, when operated out of resonance, the output sensitivity will strongly depend on the distance between the wires, because it affects the demagnetization factor. A detailed study on the core's size dependence of orthogonal fluxgates' parameters will be also useful to optimize the geometry of micro-fluxgates, where the small dimension strongly affects the achieved sensitivity and noise.

Finally, further steps should be made towards developing manufacturing techniques for the production of magnetic wires to be used as the core of orthogonal fluxgates, as a means of assuring mass production of cores with very similar parameters. Such efforts are an important requirement for the industrialization of this type of sensor.

10. Acknowledgment

The author thanks the Japanese Society for Promotion of Science (JSPS) for support under the framework of the JSPS PostDoc fellowship program. This work was supported by a kakenhi grant 22 · 00376.

11. References

Print books

Jiles, D. (1991). *Introduction to Magnetism and Magnetic Materials*. Chapman & Hall, ISBN: 0-412-38640-2, London

Edited books

Ripka, P. (Ed.). (2001). *Magnetic Sensors and Magnetometers*. Artech House., ISBN: 1580530575, Norwood, MA

Knobel, M.; Vázquez, M. & Kraus, L. (2003). *Handbook of Magnetic Materials*, Vol. 15, pp. 497-563

Patents

Alldredge, L.R. (1958). Magnetometer. *US Patent 2,856,581.* (October 1958)

Schonstedt, E.O. (1959). Saturable measuring device and magnetic core therefor. *US Patent 2,916,696.* (December 1959)

Takeuchi, S. ; Ichioka, S. ; Shirosaka, M. & Inoue, T. (1977). Thin film magnetometer using an orthogonal fluxgate. *US Patent 4,007,417.* (February 1977)

Papers in journals

Antonov, A.S. ; Buznikov, N.A. ; Iakubov, I.T. ; Lagarkov, A.N. & Rakhmanov, A.L. (2001). Nonlinear magnetization reversal of Co-based amorphous microwires induced by an ac current. *Journal of Physics D-Applied Physics*, Vol. 34, Iss. 5, (March 2001), pp. 752-757, ISSN 0022-3727

Atalay, S.; Ripka, P. & Bayri, N. (2010). Coil-less fluxgate effect in $(Co_{0.94}Fe_{0.06})_{72.5}Si_{12.5}B_{15}$ amorphous wires. *Journal of Magnetism and Magnetic Materials*, Vol. 322, Iss. 15, pp. 2238-2243, ISSN: 0304-8853 (August 2010)

Atalay, S.; Yagmur, V.; Atalay, F.E. & Bayri, N. (2011). Coil-less fluxgate effect in CoNiFe/Cu wire electrodeposited under torsion. Journal of Magnetism and Magnetic Materials, Vol. 323, Iss. 22, pp. 2818-2822 (November 2011)

Butta, M.; Ripka, P.; Atalay, S.; Atalay, F.E. & Li, X.P. (2008a). Fluxgate effect in twisted magnetic wire. *Journal of Magnetism and Magnetic Materials,* Vol. 320, Iss. 20, pp. e974-e978, ISSN: 0304-8853

Butta, M. & Ripka, P. (2008b). Two-domain model for orthogonal fluxgate. *IEEE Transactions on Magnetics.* Vol. 44, Iss. 11, pp. 3992-3995, ISSN: 0018-9464 (November, 2008)

Butta, M.; Ripka, P.; Infante, G.; Badini Confalonieri, G.A. & Vázquez, M. (2009a). Bi-metallic magnetic wire with insulating layer as core for orthogonal fluxgate. *IEEE Transactions on Magnetics*, Vol. 45, Iss. 10, pp. 4443-4446, ISSN: 0018-9464

Butta, M. & Ripka, P. (2009b). Model for coil-less fluxgate. *Sensors and Actuators A: Physical,* Vol. 156, Iss. 1, pp. 269-273. ISSN: 0924-4247 (November 2009)

Butta, M; Ripka, P.; Perez Navarrete, J. & Vázquez, M. (2010a). Double coil-less fluxgate in Bridge configuration. *IEEE Transactions on Magnetics*, Vol. 46, Iss. 2, pp. 532-535, ISSN: 0018-9464

Butta M.; Ripka, P.; Infante, G.; Badini-Confalonieri G.A. & Vázquez, M. (2010b). Magnetic microwires with field induced helical anisotropy for coil-less fluxgate. *IEEE Transactions on Magnetics,* Vol. 46, Iss. 7, pp. 2562-2565, ISSN: 0018-9464

Butta M., Yamashita, S. & Sasada, I. (2011). Reduction of noise in fundamental mode orthogonal fluxgate by optimization of excitation current. Accepted in *IEEE Transactions on Magnetics*

Gise, P.E. & Yarbrough, R.B. (1975). An electrodeposited cylindrical magnetometer sensor. *IEEE Transactions on Magnetics*, Vol. 11, Iss. 5, (September 1975), pp. 1403-1404, ISSN: 0018-9464

Gise, P.E. & Yarbrough, R.B. (1977). An improved cylindrical magnetometer sensor. *IEEE Transactions on Magnetics.* Vol. 13, Iss. 5, (September 1977), pp. 1104-1106, ISSN 0018-9464

Janosek, M. & Ripka, P. (2009). Current-output of printed circuit board fluxgates. *Sensor letters*, Vol. 7, Iss. 3, pp.299-302, ISSN: 1546-198X (June 2009)

Jie F.; Li, X.P. & Ripka, P. (2006). Low power orthogonal fluxgate sensor with electroplated Ni80Fe20/Cu wire. *Journal of Applied Physiscs*, Vol. 99, Iss. 8 (April 2006), Article number 08B311, ISSN 0021-8979

Jie, F. ; Ning, N. ; Ji, W. ; Chiriac, H. ; Li, X.P. (2009). Study of the Noise in Multicore Orthogonal Fluxgate Sensors Based on Ni-Fe/Cu Composite Microwire Arrays. *IEEE Transaction on Magnetics*, Vol. 45, Iss. Sp. Iss. SI, (October 2009), pp. 4451-4454, ISSN 0018-9464

Kubik, J.; Janosek, M. & Ripka, P. (2007). Low-power fluxgate sensor signal processing using gated differential integrator. *Sensors Letters*, Vol. 5, Iss. 1, pp. 149-152, ISSN : 1546-198X

Kubik, J.; Pavel, L.; Ripka, L. & Kaspar, P. (2007). Low-Power Printed Circuit Board Fluxgate Sensor. *IEEE Sensors Journal*, Vol. 7, Iss. 2, pp. 179-183

Kubik, J.; Vcelak, J.; O'Donnel, T. & McCloskey, P. (2009). Triaxial fluxgate sensor with electroplated core. *Sensors and Actuators A: Physical*, Vol. 152, Iss. 2, pp. 139-145, ISSN: 0924-4247 (June 2009)

Li, X.P. ; Zhao, Z.J. ; Seet, H.L. ; Heng, W.M. ; Oh, T.B. & Lee, J.Y. (2003).Effect magnetic field on the magnetic properties of electroplated NiFe/Cu composite wires. *Journal of Applied Physics*. Vol. 94, Iss. 10, (November 2003), pp. 6655-6658, ISSN 0021-8979

Li, X.P. ; Zhao Z.J. ; Oh, T.B. ; Seet, H.L. ; Neo, B.H. & Koh S.J. (2004). Current driven magnetic permeability interference sensor using NiFe/Cu composite wire with a signal pick-up LC circuit. *Physica status solidi A-Applied Research*, Vol. 201, Iss. 8, (June 2004), pp. 1992-1995, ISSN 0031-8965

Li, X.P. ; Fan, J. ; Ding, J.; Chiriac, H.; Qian, X.B. & Yi, J.B. (2006a). A design of orthogonal fluxgate sensor, *Journal of Applied Physics*, Vol. 99, Iss. 8, (April 2006), Article number 08B313, ISSN 0021-8979

Li, X.P. ; Fan, J. ; Ding, J. & Qian, X.B. (2006b). Multi-core orthogonal fluxgate sensor. *Journal of Magnetism and Magnetic Materials*, Vol. 300, Iss. 1 (May 2006), pp. e98-e103

Li, X.P. ; Seet, H.L. ; Fan, J. & Yi, J.B. (2006c). Electrodeposition and characteristics of Ni80Fe20/Cu composite wires. Journal of Magnetism and Magnetic Materials, Vol. 304, Iss. 1 (September 2006), pp. 111-116, ISSN 0304-8853

Malatek, M., Ripka, P. & Kraus, L. (2005). Double-core GMI current sensor. *IEEE Transactions on Magnetics*, Vol. 41, Iss. 10, pp. 3703-3705, ISSN: 0018-9464 (October 2005)

Paperno, E. (2004). Suppression of magnetic noise in the fundamental-mode orthogonal fluxgate. *Sensors and Actuators A-Physical*, Vol. 116, Iss. 3, (October 2004), pp. 405-409, ISSN 0924-4247

Paperno, E. ; Weiss, E. & Plotkin, A. (2008). A Tube-Core Orthogonal Fluxgate Operated in Fundamental Mode. *IEEE Transaction on Magnetics*, Vol. 44, Iss. 11 (November 2008), pp. 4018-021, ISSN 0018-9464

Primdahl, F. (1970). The fluxgate mechanism, Part I : The gating curves of parallel and orthogonal fluxgates. *IEEE Transactions on Magnetics*, Vol. MAG-6, Iss. 2 (June 1970), pp. 376-383, ISSN 0018-9464

Ripka, P.; Li, X.P. & Jie, F. (2005). Orthogonal fluxgate effect in electroplated wires. *2005 IEEE Sensors*

Ripka, P.; Butta, M.; Malatek, M.; Atalay, S. & Atalay, F.E. (2008). Characterization of magnetic wires for flugate cores. *Sensors and Actuators A – Physical*, Vol. 145-146, Special Issue, pp. 23-28, ISSN: 0924-4247 (July-August, 2008)

Ripka, P.; Li, X.P. & Jie, F. (2009). Multiwire core fluxgate. *Sensors and Actuators A - Physical*, Vol. 156, Iss.1 (November 2009), pp. 265-268, ISSN 0924-4247

Ripka, P. ; Butta, M., Jie, F. & Li, X.P. (2010). Sensitivity and Noise of Wire-Core Transverse Fluxgate. *IEEE Transactions on Magnetics*, Vol. 46, Iss. 2 (February 2010), pp. 654-657, ISSN 0018-9464

Sasada, I. (2002a). Orthogonal fluxgate mechanism operated with dc biased excitation. *Journal of Applied Physiscs*, Vol. 91, Iss. 10, (May 2002), pp. 7789-7791, ISSN 0021-8979

Sasada, I. (2002b). Symmetric response obtained with an orthogonal fluxgate operating in fundamental mode. *IEEE Transactions on magnetics*, Vol. 38, Iss. 5, (September 2002), pp. 3377-3379, ISSN 0018-9464

Sasada, I. & Kashima, H. (2009) Simple Design for Orthogonal Fluxgate Magnetometer in Fundamental Mode. *Journal of the Magnetics Society of Japan*, Vol. (33), n. 2, (2009), pp. 43-45, ISSN 1882-2924

Seet, H.L.; Li, X.P.; Ning, N. ; Ng, W.C. & Yi, J.B. (2006). Effect of magnetic coating layer thickness on the magnetic properties of electrodeposited NiFe/Cu composite wires. *IEEE Transactions on magnetics*, Vol. 42, Iss. 10, (October 2006), pp. 2784-2786, ISSN 0018-9464

Sinnecker, J.P.; Pirota, K.R.; Knobel, M. & Kraus, L. (2002). AC magnetic transport on heterogeneous ferromagnetic wires and tube. *Journal of Magnetism and Magnetic Materials*, Vol. 249, Iss. 1-2, pp. 16-21 (2002)

Terashima, Y. & Sasada, I. (2002). Magnetic domain Imaging using orthogonal fluxgate probes. *Journal of Applied Physics*, Vol. 91, Iss. 10 (May 2002), pp. 8888-8890, ISSN 0021-8979

Vázquez, M. & Hernando, A. (1996). A soft magnetic wire for sensor applications. *Journal of Physics D: Apllied Physics*, Vol. 29, Iss. 4, pp. 939-949, ISSN: 0022-3727 (April 1996)

Vázquez, M.; Chiriac, H.; Zhukov, A.; Panina, L. & Uchiyama, T. (2011). On the state-of-the-art in magnetic microwires and expected trends for scientific and technological studies. *Physica Status Solidi (a)*, Vol. 208, Iss. 3, pp. 493-501, ISSN: 1862-6300 (March 2011)

Weiss, E. ; Paperno, E. & Plotkin, A. (2010). Orthogonal Fluxgate employing discontinuous excitation. *Journal of Applied Physics*, Vol. 107, Iss. 9, Article number 09E717, ISSN : 0021-8979

Zorlu, O.; Kejik, P., Vincent, F. & Popovic, R.S. (2005). A novel planar magnetic sensor based on orthogonal fluxgate principle. *Procedings of International Conference on PhD Research in Microelectronics and Electronics (PRIME 2005)*, pp. 215-218, ISBN 0-7803-9345, Lausanne, Switzerland

Zorlu, O.; Kejik, P. & Popovic, R.S. (2006). An orthogonal fluxgate-type magnetic microsensor with electroplated Permalloy core. *Sensors and Actuators A-Physical*, Vol. 135, Iss. 1, (March 2007), pp. 43-49, ISSN 0924-4247

Papers in conference proceedings

Kraus, L.; Butta, M. & Ripka P. (2010). Magnetic anisotropy and giant magnetoimpedance in NiFe electroplated on Cu wires. *Proceedings of EMSA 2010*, Bodrum, July 2010

Butta, M; Ripka, P.; Vazquez, M.; Infante, G. & L. Kraus (2010c). Microwire electroplated under torsion as core for coil-less fluxgate. Proceedings of EMSA 2010, Bodrum, July 2010

2

Analysis of Thermodynamic and Phononic Properties of Perovskite Manganites

Renu Choithrani
Department of Physics, Barkatullah University, Bhopal
India

1. Introduction

The perovskite manganites are of recognized importance, owing to the development of new materials designed for different potential scientific and technological applications, such as *magnetic sensors in the tracking and location of emergency personnel, ferroelectric thin films, microwaves, spintronics, semiconductor technologies and memory devices* etc. These materials have been the subject of significant research interest because of the intriguing underlying physics showing marked colossal magnetoresistance (CMR) effect and the anticipated multifunctional and advanced applications for the next generation electronics. The recent discovery of large magnetoelectric effects in the $R_{1-x}A_x\text{MnO}_3$ (where R and A are trivalent rare earth and divalent alkaline earth ions, respectively) has kindled interest among investigators in understanding the complex relationships between lattice distortion, magnetism, dielectric and thermodynamic properties of undoped $R\text{MnO}_3$ and are good candidates for certain sensor applications, bolometers, magnetic refrigeration, read head devices, magnetic storage in hard disk and floppy disks, and spin valve devices (such as the 21st Century electronics i.e. spintronics (Yamasaki et al., 2007). The huge magnetoresistance or colossal magnetoresistance effect (CMR) has been observed in the perovskite manganites (Jin et al., 1994). The *magnetoresistance* is defined as ($\Delta R/R_H = [R(T,H) - R(T, 0)]/R(T, H)$. Magnetoresistance can be negative or positive. In most nonmagnetic solids the magnetoresistance is positive. In non-magnetic pure metals and alloys MR is generally positive and MR shows a quadratic dependence on H. MR can be negative in magnetic materials because of the suppression of "Spin Disorder" by the magnetic field. The Giant magnetoresistive effect GMR was recently discovered by Sir Peter Grünsberg (Germany) and Sir Albert Flert (France) (see photographs 1 and 2) and both of them were awarded Nobel Prize in the year 2007 (Baibich et al., 1988; Binasch et al., 1989). It has been used extensively in read heads of modern hard drives. Another application of the GMR effect is non-volatile, magnetic random access memory. Read heads of modern hard drives are based on the GMR or TMR effect.The GMR effect is observed, mainly, in artificial multilayer systems with alternating magnetic structure and is used in many applications. The large magnetoelectric effects in the rare-earth manganites $R\text{MnO}_3$ has reopened the field of the so-called *multiferroic materials*. Multiferroic means, that several ferro-type orders, such as ferromagnetism, ferroelectricity or ferroelasticity coexist.

1   2

Photograph 1. Peter Grünberg, 2. Albert Fert Nobel Prize Winners of 2007

Manganites have opened up challenging and exciting possibilities for research in condensed matter physics and materials science. Today, the science and technology of magnetism have undergone a renaissance, driven both by the urge to understand the new physics and by demand from industries for better materials. The current computer age relies on the development of smaller and "smaller" materials for memory devices, data storage, processing and probing. Materials can be divided into numerous categories, depending on their origin, use or morphology. The number of applications, properties and their combinations seems to be almost unlimited. One prime example of a small but diverse group of materials is the perovskite family. Solid state physicists, chemists and materials scientists have recently shown a wide interest in these perovskites, because of their unique properties such as the high temperature superconductivity, colossal magnetoresistance (CMR) and ferroelectricity. Still new and unexplored possibilities of the perovskites appear. In spite of the progress made in our understanding of the properties of these materials, very basic questions like "Why is this material a good conductor, while a similar material is an insulator" remain a challenge. Millions of people tackle this question everyday. Manganites clearly offer great opportunity at low temperatures but extending this promise to room temperature remains a challenge.

Despite enormous theoretical and experimental activity in the area of manganites, a complete understanding of the unusual physical properties in perovskite manganites is still lacking and many questions remain unresolved. For example, a lot of theoretical models are developed, but they are so diverse that it is difficult to choose between them. Moreover, a quantitative comparison of the known models with experimental data is practically impossible or too ambiguous. After several theoretical efforts in recent years, mainly guided by the computational and mean-field studies of realistic models, considerable progress has been achieved in understanding the physical properties of these compounds. Most vividly, to the best of my knowledge, only a few groups are involved with the theoretical investigation of the temperature, pressure and composition-dependent thermodynamical, structural and phonon dynamical (vibrational) properties of perovskite manganites and/or multiferroic manganites.

Recently, a modified rigid ion model (MRIM) has been developed by the author (Renu Choithrani et al., 2008-2010) to elucidate the cohesive, thermal and thermodynamic properties of pure and doped perovskite manganites and the study is recognized in journals of international repute with high impact factor, awarded and appreciated by the eminent scientists, Nobel Laureates, reputed national and international Universities, Research Centres and R & D groups.

The system offers a degree of chemical flexibility which permits the relation between the structural, electronic and magnetic properties of these oxides to be examined in a systematic way. The manganites also have potential as solid electrolytes, catalysts, sensors and novel electronic materials. Their rich electronic phase diagrams reflect the fine balance of interactions which determine the electronic ground state. These compounds represent the interplay of experiment, theory and application which is an important aspect of condensed matter physics research.

The experiments have given quantitative explanation of the metal-insulator phase transitions, while from a fundamental point of view, the central question is to understand the different physical properties and thus, the theoretical attempts are still awaited to understand the mechanism of these exotic materials. Looking to the importance of materials and availability of wealth of experimental data on various physical properties. It is well known that these physical properties of the materials originate from the interatomic interactions between their atoms. Therefore, the author has thought it pertinent to develop the model of interatomic potentials with limited number of adjustable parameters to predict the cohesive, thermal, elastic and phononic dynamical properties of the pure and doped perovskites. The perovskite structure is adopted by many oxides that have the chemical formula ABO_3. The relative ion size requirements for stability of the cubic structure are quite stringent, so slight buckling and distortion can produce several lower-symmetry distorted versions, in which the coordination numbers of A , B ions or both are reduced. The stability of the perovskite structure depends upon the tolerance factor defined by Goldsmith tolerance factor (t). The perovskite structure is supposed to be stabilized in the range of 0.75< t < 1. However, for an ideal perovskite t = 1. For perovskite containing small central cations, the Goldschmidt tolerance factor, t, is typically <1. Tilting of the octahedral structure is common in perovskite with t<1. For t<1, the cubic structure transforms to the orthorhombic structure which leads to deviation in the Mn-O-Mn bond angle from 180° (for an ideal perovskite). The deviation of the Mn-O-Mn bond angle from 180° leads to the distortion in the MnO_6 octahedron. According to the Jahn-Teller (JT) theorem, the structure will distort thereby removing the degeneracy of the e_g orbitals. In solids, the orbital degree of freedom of the Mn^{3+} ion often shows long range ordering associated with the cooperative JT effect. The magnetic properties of the manganites are governed by exchange interactions between the Mn ion spins. These interactions are relatively large between two Mn spins separated by an oxygen atom and are controlled by the overlap between the Mn d-orbitals and O p-orbitals.

The present investigations provide several important information about their cohesive, thermal, elastic, phonon and vibrational properties and how these properties of pure and doped perovskite manganites vary at magnetic transitions. The learning of inter molecular interaction is important as it influences the stability and hardness of the material. It is also distinguished that the chemical stability is associated with cohesive energy. Compounds with higher cohesive energy generally have chemical stability. The study of atomic vibrations in crystalline solids is a subject with a long and interesting history and over the years has been explored in all its aspects. The formal basis for the investigation of the vibrations, especially within the framework of the harmonic approximation, along with meritorious treatments in their own right has been presented in this chapter. The modified effective interaction potentials to explore various physical properties of these advanced, smart and novel materials are described in detail in this chapter.

2. Importance

The development of new and novel materials for technological applications has opened many doors to innovation in the 21st century. New magnetic and electronic materials in particular have helped to bring about the information revolution. Many physicists are interested in new materials because they can be used to study a new physical phenomenon. With respect to structure, these manganites have been grouped into the hexagonal phase (*P63cm*) with R (= Ho, Er, Tm, Yb, Lu or Y) (Yakel et al., 1963) having smaller ionic radius (r_R) and the orthorhombic structure (*Pnma*) with R (= La, Pr, Nd, Sm, Eu, Gd, Tb or Dy) (Gilleo et al., 1957) having larger (r_R)(figures 1 and 2).

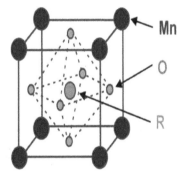

Fig. 1. Crystal lattice of the orthorhombic manganites $RMnO_3$.

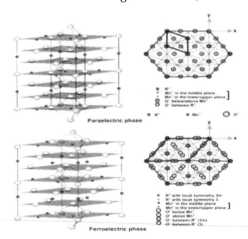

Fig. 2. Crystal lattice of the hexagonal manganites $RMnO_3$.

Research work on perovskite materials is being carried out both in India and other countries and has resulted into several important technological applications and patents, particularly in the areas like spintronics, semiconductor technologies, magnetic sensors, bolometers, magnetic refrigeration, read head devices, magnetic storage in hard disk and floppy disks. At the international level, an extensive research work is being conducted by several groups (Alonso et al., 2000; Lo´pez et al., 2002; Ku et al., 2004; Kimura et al., 2005; Iliev et al., 2006; Tachibana et al., 2007;) on pure and doped perovskite manganites. The research work in India is being conducted mostly at Indian Institute of Science, Bangalore (Prof. C.N.R. Rao and collaborators, 2001), Indian Institute of Technology, Delhi (Prof. H.C. Gupta and collaborators, 2008) Bhabha Atomic Research Centre, Mumbai (Prof. S.L. Chaplot and collaborators, 2011), and at several other organizations. Less theoretical attempts have been made to understand the mechanism. Interesting peculiar features are observed due to the effect of doping of rare earth (Ca, Sr, Ba,.....) in $RMnO_3$. Less attention has been paid on subterranean understanding of thermodynamic and phononic properties.

3. Detailed methodology

3.1 Extended Rigid Ion Model (ERIM)

The author has recently developed an extended rigid ion model (ERIM) by incorporating the the long-range (LR) Coulomb attraction, the short-range (SR) Hafemeister–Flygare (HF) type overlap repulsion effective up to the second neighbour ions, the van der Waals (vdW) attraction due to the dipole–dipole (d–d) and dipole–quadrupole (d–q) interactions and zero point energy (ZPE) effects in the framework of modified rigid ion model (MRIM) developed earlier by us(Renu Choithrani et al., 2008-2010).

The framework of ERIM is derived from the following interionic interaction potential:

$$\phi_{ERIM} = \phi_{MRIM} + \phi_{ZPE}$$

where, ϕ_{MRIM} potential (Renu Choithrani et al., 2008-2010) is given by

$$\phi_{MRIM} = -\frac{e^2}{2}\sum_{kk'} Z_k Z_{k'} r_{kk'}^{-1} + \begin{bmatrix} n b_1 \beta_{kk'}\exp\{(r_k + r_{k'} - r_{kk'})/\rho_1\} + \\ \frac{n'}{2}b_2\left[\beta_{kk}\exp\{(2r_k - r_{kk})/\rho_2\} + \beta_{k'k'}\exp\{(2r_{k'} - r_{k'k'})/\rho_2\}\right] \end{bmatrix} - \sum_{kk'}c_{kk'}r_{kk'}^{-6} - \sum_{kk'}d_{kk'}r_{kk'}^{-8} \quad (1)$$

and

$$\phi_{ZPE} = (9/4)\,K\theta_D \quad (2)$$

This ERIM has been developed and applied by the author, probably for the first time, to describe the thermodynamic and phononic properties of the doped and undoped perovskites materials.

$\beta_{kk'}$ are the Pauling coefficients:

$$\beta_{kk'} = 1 + (z_k / n_k) + (z_{k'} / n_{k'}) \tag{3}$$

with z_k ($z_{k'}$) and n_k ($n_{k'}$) as the valence and number of electrons in the outermost orbits of k (k') ions. In Eqn. (1), $r_{kk'}$ and r_{kk}(= $r_{k'k'}$) are, respectively, the first and second neighbour separations and their values are obtained for A substitutions (x) using the well known Vegard's law (Vegard, 1921).

The symbols c_{kk}' and d_{kk}' are the corresponding vdW coefficients, whose values have been determined by using their expressions derived from the Slater-Kirkwood variational (SKV) (Slater & Kirkwood, 1931) method:

$$c_{kk'} = (3e\,\hbar\,\alpha_k\,\alpha_{k'} / 2m)\,[(\alpha_{k/Nk})^{1/2} + (\alpha_{k'/Nk'})^{1/2}]^{-1} \tag{4}$$

$$d_{kk'} = (27e\,\hbar^2\alpha_k\,\alpha_{k'} / 8m)\,[(\alpha_{k/Nk})^{1/2} + (\alpha_{k'/Nk'})^{1/2}]^2 \tag{5}$$

$$[(\alpha_{k/Nk})^{1/2} + 20/3\,(\alpha_k\,\alpha_{k'/\,Nk\,Nk'})\,(\alpha_{k'/Nk'})]^{-1}$$

where m and e are the mass and charge of electron, respectively. $a_k(a_{k'})$ are the electronic polarizabilities of k(k') ions ; N_k ($N_{k'}$) are the effective number of electrons responsible for the polarization of k(k') ions. The values of $c_{kk'}$ and $d_{kk'}$ are evaluated using the Eqns. (4) and (5) and the procedure prescribed in our earlier papers (Renu Choithrani et al., 2008-2010).

The model parameters (hardness (b) and range (ρ)) are determined from the equilibrium condition:

$$[d\phi(r) / dr]_{r=r_0} = 0 \tag{6}$$

and the Bulk modulus:

$$B = (9Kr_0)^{-1}[d^2\phi(r) / dr^2]_{r=r_0} \tag{7}$$

here, r_0 and r are the interionic separations in the equilibrium and otherwise states of the system, respectively. The symbol K is the crystal structure constant.

The total specific heat is computed by including the contribution of specific   heat from lattice, electrons, phonons and schottky effects.

$$C_{lat} = 9RN(\frac{T}{\theta_D})\int_0^{\theta_D/T} \frac{x^4 e^x}{(e^x - 1)^2} dx \tag{8}$$

where N=5 is the number of atoms in the unit cell, R = 8.314 J /mol K is the ideal gas constant, θ_D is the Debye temperature and x = θ_D/T.

$$C_p = C_e + C_{ph} + C_{sch} \tag{9}$$

$$C_e = \gamma T \tag{10}$$

$$C_{Ph} = \beta T^3 \tag{11}$$

$$C_{sch} = \frac{R}{T^2}\left\{\sum_{i=1}^{n}\Delta_i^2\exp(\frac{-\Delta_i}{T})/\sum_{i=1}^{n}\exp(\frac{-\Delta_i}{T}) - \left[\sum_{i=1}^{n}\Delta_i\exp(\frac{-\Delta_i}{T})/\sum_{i=1}^{n}\exp(\frac{-\Delta_i}{T})\right]^2\right\} \tag{12}$$

The computations have been performed to understand the cohesive and thermal properties at different temperatures (T) and compositions (x) of $RMn_{1-x}Ga_xO_3$ (R = Ho, Y and x =0, 0.03, 0.1, 0.2 and 0.3). Here, the author has tried to address the study of thermodynamic properties of solids which is one of the fascinating fields of condensed matter physics in the recent years. Thermodynamics play an important role in explaining the behaviour of manganites, as many properties are attributed to 'electron–phonon' interaction. As electron–phonon interaction is one of the most relevant contributions in determining the conduction mechanism in these materials. The strong coupling between the electron and the lattice brings about by a small change in the chemical composition, like the ratio between trivalent and divalent ions at the A site or the average ionic radius of the ions on the A site, a large changes in the physical properties. Thus, the thermodynamic and phononic properties of these materials may be taken as a starting point to a consistent understanding of the more complex physical properties of the manganites. The author has explored the physical properties such as cohesive energy (ϕ), molecular force constant (f), compressibility (β), Restrahalen frequency (υ_0), Debye temperature (θ_D), Grüneisen parameter (γ) and specific heat (C(T)), probably for the first time, using extended rigid ion model (ERIM).

4. Results and discussion

Using the input data (Zhou et al., 2006) and the vdW coefficients (c_{kk}' and d_{kk}') calculated from the SKV method, the model parameters ((b_1, ρ_1) and (b_2, ρ_2)) corresponding to the ionic bonds Mn-O and R/Ga–O for different compositions ($0.0 \leq x \leq 0.3$) and temperatures 0 K $\leq T \leq$ 300 K have been calculated using equations (6) and (7) for $RMn_{1-x}Ga_xO_3$ (R = Ho, Y and x = 0, 0.03, 0.1, 0.2 and 0.3). Using the disposable model parameters listed in tables 1 and 2, the values of ϕ, f , β, υ_0, θ_D and γ are computed and depicted in tables 3 and 4 for $RMn_{1-x}Ga_xO_3$ (R = Ho, Y and x = 0, 0.03, 0.1, 0.2 and 0.3), respectively. The chief aim of the application of the present model is to reproduce the observed physical properties such as the phononic and thermodynamic properties of perovskite manganites. Keeping this objective in view, we find that our results obtained for most of these properties are closer to the experimental data (Zhou et al., 2006) available only at 300 K. The cohesive energy is the measure of strength of the force binding the atoms together in solids. This fact is exhibited from cohesive energy results which follow the similar trend of variation with T, as is revealed by the bulk modulus that represents the resistance to volume change. This feature is indicated from tables 3 and 4, which show that the values of the cohesive energy (ϕ) decreases from −148.01 eV for $HoMn_{1-x}Ga_xO_3$ (x = 0) to −149.52 eV for $HoMn_{1-x}Ga_xO_3$ (x = 0.2) and −146.99 eV for $YMn_{1-x}Ga_xO_3$ (x = 0) to −148.87 eV for $YMn_{1-x}Ga_xO_3$ (x = 0.3) dopings; the same trend of variation is also exhibited by the bulk modulus. Due to a lack of experimental data, the values of cohesive energy of $RMn_{1-x}Ga_xO_3$ (R = Ho, Y and x =0, 0.03, 0.1, 0.2 and 0.3) are compared with that of $LaMnO_3$ (De Souza et al., 1999), which is a member of the same family. The negative values of cohesive energy show that the stability of these manganites is intact. It is also noticed from tables 1 and 2 that the values of hardness (b_1, b_2) and range parameters (ρ_1, ρ_2) increase with concentrations (x). The author has also calculated the values of the molecular force constants (f) and Restrahalen frequencies (υ_0) (see tables 3 and 4) and found that υ_0 increases with temperatures (T) and concentrations (x). Since the Restrahalen frequency is directly proportional to the molecular force constant (f) therefore both of them vary with the temperature accordingly for different concentrations (x). The values of Restrahalen frequency are almost in the same range as that reported for the pure material $YMnO_3$ (υ_0 = 13.08 THz) (Zhou et al., 2006). The calculated value of Debye temperature (θ_D) for $RMn_{1-x}Ga_xO_3$ (R = Ho, Y and x =0, 0.03, 0.1, 0.2 and 0.3) is approximately in agreement and comparable with the corresponding data (577 K and 628 K) available only at room temperature for $HoMnO_3$ and $YMnO_3$ (Zhou et al., 2006; Renu Choithrani et al., 2011). The calculated values of (θ_D) for $RMn_{1-x}Ga_xO_3$ (R = Ho, Y and x =0, 0.03, 0.1, 0.2 and 0.3) at 300 K are ranging from 577.11 K to 622.05 K, which lie within the Debye temperature range (300-650 K) often found in perovskite manganites (Renu Choithrani et al., 2008-2010). The Grüneisen parameters (γ) obtained by the author are found to lie between 2 and 3, which are similar to the values observed by Dai et al., 1996.

The specific heat (C) values calculated by the author for $RMn_{1-x}Ga_xO_3$ (R = Ho, Y and x =0, 0.03, 0.1, 0.2 and 0.3) at temperatures 0 K $\leq$ T $\leq$ 300 K and are displayed in figures 3-11 and found to be in good agreement with the measured data (Zhou et al., 2006). The computed results have followed a trend more or less similar to those exhibited by the experimental curve at lower temperatures. A sharp peak is observed in the experimental specific heat curve at ~59 K due to the A-type AF ordering. The change in Mn–O distance by the substitution of a Ga increases θ_D and hence there is a consistent decrease in the specific heat values corresponding to the doping in both of the cases (see figures 3-11). Hence, the concentration (x) dependence of θ_D in $RMn_{1-x}Ga_xO_3$ (R = Ho, Y and x =0, 0.03, 0.1, 0.2 and 0.3) suggests that increased doping drives the system effectively towards the strong electron–phonon coupling regime. The present ERIM calculations yield similar specific heat values at low temperatures (below 60 K) where the acoustic phonons play an important role. On doping Ga (x = 0, 0.03, 0.1 and 0.2) in $HoMnO_3$, the specific heat increases monotonically with temperature as shown in figures 3-11. The experimental specific heat results (Zhou et al., 2006) and the computed values by ERIM for $HoMn_{1-x}Ga_xO_3$ (x =0, 0.03, 0.1 and 0.2) show three anomalies: (i) a λ-type anomaly at T_N , below 65 K(ii) a narrow peak around T_{SR}, , below 40 K and (iii) a sharp peak at T_2, below 5 K. With increasing the concentration (x), T_N and T_2 decrease but T_{SR} increases. It is also seen from figures 3-6 that the theoretical results obtained by ERIM are in closer agreement with the available experimental (Zhou et al., 2006) data. It is noticed from figures 7-11, the specific heat of $YMn_{1-x}Ga_xO_3$ just shows a λ-type peak at T_N, which decreases with increasing x. For the $YMn_{1-x}Ga_xO_3$ (x =0, 0.03, 0.1, 0.2 and 0.3) materials, the specific heat curve shows no peak with temperature down to 2 K. Another noteworthy feature is that there are broad peaks between 5 K and 10 K for $HoMn_{1-x}Ga_xO_3$, but no such peaks for $YMn_{1-x}Ga_xO_3$. The ERIM results have fairly well reproduced the experimental specific heat data in the temperature ranges 100 K $\leq$ T $\leq$ 250 K except for the range 250 K $\leq$ T $\leq$ 300 K, in which the experimental specific heat values are not available (see figures 3-11); this feature corresponds to the phase transition from a second-order nature of the antiferromagnetic (AFM) transition of the Mn^{3+} magnetic moments as revealed from the specific heat measurements were performed on a PPMS (Physical Property Measurement System, Quantum Design) data (Zhou et al., 2006).

5. Tables and figures

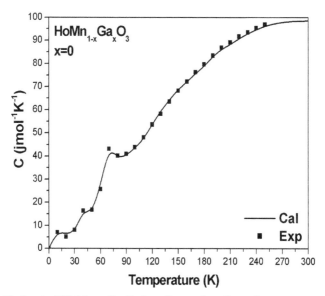

Fig. 3. The specific heat of $HoMn_{1-x}Ga_xO_3$ (x = 0) as a function of temperature.

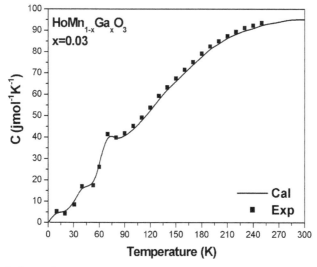

Fig. 4. The specific heat of $HoMn_{1-x}Ga_xO_3$ (x = 0.03) as a function of temperature.

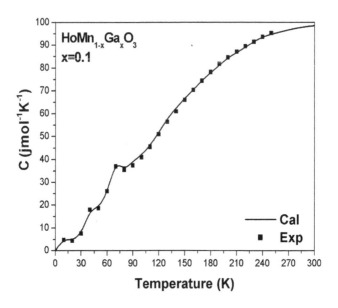

Fig. 5. The specific heat of HoMn$_{1-x}$Ga$_x$O$_3$ (x = 0.1) as a function of temperature.

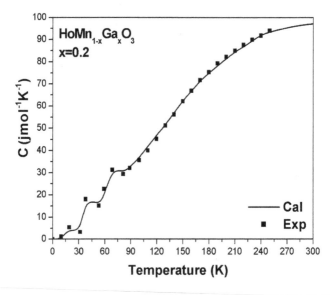

Fig. 6. The specific heat of HoMn$_{1-x}$Ga$_x$O$_3$ (x = 0.2) as a function of temperature.

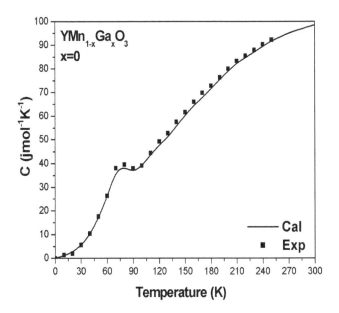

Fig. 7. The specific heat of YMn$_{1-x}$Ga$_x$O$_3$ (x = 0) as a function of temperature.

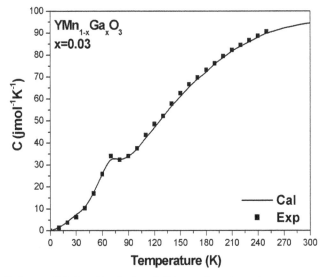

Fig. 8. The specific heat of YMn$_{1-x}$Ga$_x$O$_3$ (x = 0.03) as a function of temperature.

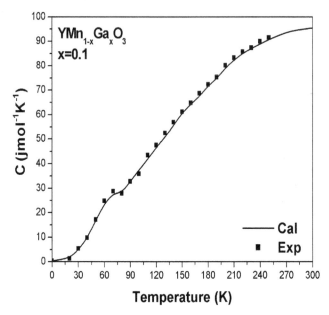

Fig. 9. The specific heat of YMn$_{1-x}$Ga$_x$O$_3$ (x = 0.1) as a function of temperature.

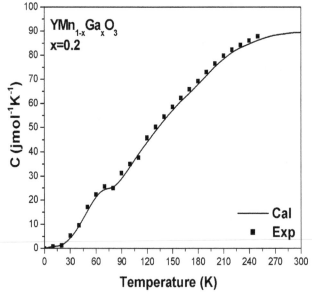

Fig. 10. The specific heat of YMn$_{1-x}$Ga$_x$O$_3$ (x = 0.2) as a function of temperature.

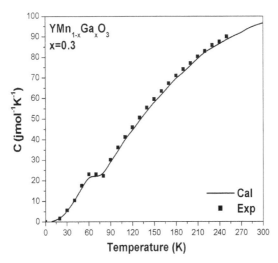

Fig. 11. The specific heat of $YMn_{1-x}Ga_xO_3$ (x = 0.3) as a function of temperature.

Concentration (x)	Model parameters			
	Mn-O		Ho/Ga-O	
	b_1 $(10^{-12}$ erg)	ρ_1 (Å)	b_2 $(10^{-12}$ erg)	ρ_2 (Å)
0	1.907	0.602	2.023	0.702
0.03	1.932	0.609	2.110	0.717
0.1	1.950	0.610	2.120	0.719
0.2	1.976	0.612	2.146	0.720

Table 1. The model parameters of $HoMn_{1-x}Ga_xO_3$ (0 ≤ x ≤ 0.2)

Concentration (x)	Model parameters			
	Mn-O		Y/Ga-O	
	b_1 $(10^{-12}$ erg)	ρ_1 (Å)	b_2 $(10^{-12}$ erg)	ρ_2 (Å)
0	1.981	0.8902	3.183	0.991
0.03	1.986	0.8922	3.197	0.994
0.1	1.991	0.8945	3.224	0.996
0.2	1.996	0.9445	3.294	0.997
0.3	2.001	0.9479	3.299	0.999

Table 2. The model parameters of $YMn_{1-x}Ga_xO_3$ (0 ≤ x ≤ 0.3)

(x)	ϕ (eV)	f $(10^4$ dyne.cm$^{-1})$	β $(10^{-12}$ dyn^{-1}.cm$^2)$	υ_0 (THz)	θ_D (K)	γ
0	-148.01	20.09	1.87	12.02	577.11	2.10
0.03	-148.42	20.76	1.89	12.06	578.99	2.12
0.1	-149.39	21.51	1.91	12.07	579.67	2.17
0.2 (Expt.)	-149.52 (-141.81)	21.97	1.99	12.10	581.12 (577)	2.19 (2-3)

Table 3. Thermophysical properties of HoMn$_{1-x}$Ga$_x$O$_3$ ($0 \leq x \leq 0.2$)

(x)	ϕ (eV)	f $(10^4$ dyne.cm$^{-1})$	β $(10^{-12}$ dyn^{-1}.cm$^2)$	υ_0 (THz)	θ_D (K)	γ
0	-146.99	24.56	2.01	12.89	618.68	2.971
0.03	-147.01	24.89	2.04	12.91	619.77	2.977
0.1	-147.59	24.97	2.17	12.92	620.24	2.982
0.2	-148.19	25.29	2.19	12.94	621.16	2.989
0.3 (Expt.)	-148.87 (-141.81)	25.99	2.21	12.95 (13.08)	622.05 (628)	2.990 (2-3)

Table 4. Thermophysical properties of YMn$_{1-x}$Ga$_x$O$_3$ ($0 \leq x \leq 0.3$)

6. Conclusion

On the basis of the overall descriptions, it may be concluded that the diverse exposition of the temperature-dependent thermodynamical and phononic properties of perovskite manganites RMn$_{1-x}$Ga$_x$O$_3$ (R = Ho, Y and x =0, 0.03, 0.1, 0.2 and 0.3) attained by the author is remarkable in view of the inherent simplicity of the extended rigid ion model. All this indicates the power and usefulness of the ERIM as having the potential to explain a variety of physical properties (such as cohesive, thermal, elastic and thermodynamic) of the pure and doped perovskite materials. However, the efforts that have been devoted by many experimental workers to observe magnetic transitions and properties, to the best of my knowledge, only a few groups are involved with the study of temperature- and composition-dependent properties of perovskite manganites (or CMR) materials. The extended rigid ion model has reproduced the physical properties that correspond well to the experimental data. The author believes that such properties will serve young researchers, scientists and experimental workers fruitfully. Also, they will get additional informations i.e. more than a thermodynamical and vibrational picture of these materials. However, it is expected to be of immense use to other workers with peripheral interest in thermodynamical and phononic. ERIM has been applied to compute the cohesive, thermal and thermodynamical properties of the pure and doped perovskites. Some of the results on these physical properties of RMn$_{1-x}$Ga$_x$O$_3$ (R = Ho, Y and x =0, 0.03, 0.1, 0.2 and 0.3) are, probably, reported for the first time. These theoretical results are of academic interest as well as of help to understand the mechanism of CMR materials. The success of the model in predicting thermodynamic and phononic properties depends crucially on their ability to explain a variety of microscopic and macroscopic dynamical properties of such complex structured materials.

7. Acknowledgments

I would like to acknowledge with thanks the Science and Engineering Research Board, Department of Science and Technology (DST), Government of India, New Delhi for providing the financial assistance and the Fast Track Young Scientist Award. It would be noteworthy to thank Dr. N.K. Gaur for his constant encouragement during the tenure of my study.

8. References

Alonso, J.A., Martı́nez-Lope, M.J., Casais, M.T., & Ferna´ndez-Dı́az, M.T. (2000). Evolution of the Jahn-Teller distortion of MnO6 octahedra in RMnO$_3$ perovskites (R) Pr, Nd, Dy, Tb, Ho, Er, Y): A neutron diffraction study, *Inorg. Chem.* Vol.(39): 917.

Baibich, M.N., Broto, J.M., Fert, A., Nguyen, F., Dau, Van, Petroff, F., Eitenne, P., Creuzet, G., Friederich, A., & Chazelas (1988). Giant magnetoresistance of (001)Fe/(001)Cr magnetic superlattices, J. *Phys. Rev. Lett.* Vol.(61): 2472.

Binasch, G., Grünberg, P., Saurenbach F., & Zinn, W. (1989). Enhanced magnetoresistance in layered magnetic structures with antiferromagnetic interlayer exchange, *Phys. Rev. B* Vol.(39): 4828.

Choithrani, Renu, & Gaur, N.K. (2008). Analysis of low temperature specific heat in Nd$_{0.5}$Sr$_{0.5}$MnO$_3$ and R$_{0.5}$Ca$_{0.5}$MnO$_3$ (R=Nd, Sm, Dy and Ho) compounds, *J. Magn. Magn. Mater.* Vol.(320): 3384.

Choithrani, Renu, & Gaur, N.K. (2008). Heat capacity of EuMnO$_3$ and Eu$_{0.7}$A$_{0.3}$MnO$_3$ (A =Ca, Sr) compounds, *J. Magn. Magn. Mater.* Vol.(320): 10.

Choithrani, Renu, Gaur, N.K., & Singh, R.K. (2008). Specific heat and transport properties of La$_{1-x}$Gd$_x$MnO$_3$ at 15 K ≤ T ≤ 300 K, *Solid. State Commun.* Vol.(147): 103.

Choithrani, Renu, Gaur, N.K., & Singh, R.K. (2008). Thermodynamic properties of SmMnO$_3$, Sm$_{0.55}$Sr$_{0.45}$MnO$_3$ and Ca$_{0.85}$Sm$_{0.15}$MnO$_3$, *J. Phys.: Condens. Matter* Vol.(20): 415201.

Choithrani, Renu, & Gaur, N.K. (2008). Thermo physical properties of multiferroic rare earth manganite GdMnO$_3$, *AIP Proc.* Vol.(1004): 73.

Choithrani, Renu, Gaur, N.K., & Singh, R.K. (2009). Study of calcium doping effect on thermophysical properties of some perovskite manganites, *J. Alloys and Compounds* Vol.(480): 727.

Choithrani, Renu, Gaur, N.K., & Singh, R.K. (2009) . Influence of temperature and composition on cohesive and thermal properties of mixed crystal manganites, *J. Magn. Magn. Mater.* Vol.(321): (2009) 4086.

Choithrani, Renu, & Gaur, N.K. (2010). Specific heat of Cd-doped manganites, *J. Comput. Mat. Sci.* Vol.(49): 107.

Choithrani, Renu, Rao, Mala N, Chaplot, S.L., Gaur, N. K., & Singh, R. K. (2011). Lattice dynamics of manganites RMnO$_3$ (R = Sm, Eu or Gd): instabilities and coexistence of orthorhombic and hexagonal phases, New *J. Phys* Vol.(11): 073041; (2009). Structural and phonon dynamical properties of perovskite manganites: (Tb,Dy, Ho)MnO$_3$, *Journal of Magnetism and Magnetic Materials* Vol.(323): 1627.

Dai, P., Jiandi, Zhang, Mook, H.A., Lion, S.H., Dowben, P.A., & Plummer, E.W. (1996). Experimental evidence for the dynamic Jahn-Teller effect in La$_{0.65}$Ca$_{0.35}$MnO$_3$, *Phys.Rev. B* Vol.(54): 3694(R).

De Souza, Roger A., Islam, S., M., & Tiffe, E.I. (1999). Formation and migration of cation defects in the perovskite oxide LaMnO$_3$, *J. Mater. Chem.* Vol.(9) : 1621.

Gilleo, M.A. (1957). Crystallographic studies of perovskite-like compounds. III. La(M_x, Mn$_{1-x}$)O$_3$ with M = Co, Fe and Cr, *Acta Crystallogr.* Vol.(10): 161.

Gupta, H.C., & Tripathi, U. (2008). Zone center phonons of the orthorhombic RMnO$_3$ (R = Pr, Eu, Tb, Dy, Ho) perovskites, *PMC Physics B* Vol.(1): 9.

Iliev, M.N., Abrashev, M.V., Laverdière, J., Jandl, S., Gospodinov, M.M., Wang, Y.-Q., & Sun, Y.-Y. (2006). Distortion-dependent Raman spectra and mode mixing in RMnO$_3$ perovskites (R=La,Pr,Nd,Sm,Eu,Gd,Tb,Dy,Ho,Y), *Phys.Rev. B* Vol.(73): 064302.

Jin, S., Tiefel, T.H., McCormack, M., Fastnacht, R.A., Ramesh, R. & Chen, L.H. (1994). Thousandfold change in resistivity in magnetoresistive La-Ca-Mn-O films, *Science* Vol.(264): 413.

Kimura, T., Lawes, G., Goto, T., Tokura, Y., & Ramirez, A.P. (2005). Magnetoelectric phase diagrams of orthorhombic RMnO$_3$ (R = Gd, Tb, and Dy), *Phys. Rev. B* Vol.(71): 224425.

Ku, H.C., Chen, C.T., & Lin, B.N. (2004). A-type antiferromagnetic order, 2D ferromagnetic fluctuation and orbital order in stoichiometric La$_{1-x}$Eu$_x$MnO$_3$, *J. Magn. Magn. Mater.* Vol.(272): 85.

Lo´pez, J., de Lima, O.F., Lisboa-Filho, P.N., & Araujo-Moreira, F.M. (2002). Specific heat at low temperatures and magnetic measurements in Nd$_{0.5}$Sr$_{0.5}$MnO$_3$ and $R_{0.5}$Ca$_{0.5}$MnO$_3$ (R=Nd, Sm, Dy, and Ho) samples, *Phys. Rev. B* Vol.(66): 214402.

Raychaudhuri, A.K., Guha, A., Das, I., Rawat, R., & Rao, C.N.R. (2001). Specific heat of single-crystalline Pr$_{0.63}$Ca$_{0.37}$MnO$_3$ in the presence of a magnetic field, *Phys. Rev. B* Vol.(64): 165111.

Slater, J.C., & Kirkwood, K.G. (1931). The van der Waals forces in gases, *Phys. Rev. Lett.* Vol.(37): 682.

Tachibana, M., Shimoyama, T., Kawaji, H., Atake, T., & Muromachi, E. T. (2007). Jahn-Teller distortion and magnetic transitions in perovskite RMnO$_3$ (R=Ho, Er, Tm, Yb, and Lu), *Phys.Rev. B* Vol.(75): 144425.

Vegard, L. (1921). The constitution of mixed crystals and filling the space of atoms, *Z. Phys.* Vol.(5): 17.

Yakel, H.L., Koehler, W.C., Bertaut, E.F., & Forrat, E.F. (1963). On the crystal structure of the manganese(III) trioxides of the heavy lanthanides and yttrium, *Acta Crystallogr.* Vol.(16): 957.

Yamasaki, Y., Miyasaka, S., Goto, T., Sagayama, H., Arima, T. & Tokura, Y. (2007). Ferroelectric phase transitions of 3d-spin origin in Eu$_{1-x}$Y$_x$MnO$_3$, *Phys. Rev. B* Vol.(76): 184418.

Zhou, H.D., Janik, J. A., Vogt, B. W., Jo, Y. J., Balicas, L., Case, M. J., Wiebe, C. R., Denyszyn, J. C., Goodenough, J. B., & Cheng, J. G. (2006). Specific heat of geometrically frustrated and multiferroic RMn$_{1-x}$Ga$_x$O$_3$ (R=Ho,Y), *Phys.Rev. B* Vol. (74): 094426.

3

Induction Magnetometers Principle, Modeling and Ways of Improvement

Christophe Coillot and Paul Leroy
LPP Laboratory of Plasma Physics
France

1. Introduction

Induction sensors (also known as search coils), because of their measuring principle, are dedicated to varying magnetic field measurement. Despite the disadvantage of their size, induction magnetometer remains indispensable in numerous fields due to their sensitivity and robustness whether for natural electromagnetic waves analysis on Earth Lichtenberger et al. (2008), geophysics studies using electromagnetic toolsHayakawa (2007); Pfaffling (2007), biomedical applications Ripka (2008) or space physic investigations Roux et al. (2008). The knowledge of physical phenomena related to induction magnetometers (induction, magnetic amplification and low noise amplification) constitutes a strong background to address design of other types of magnetometers and their application.

Let us describe the magnetic field measurement in the context of space plasma physics. AC and DC magnetic fields are among the basic measurements you have to perform when you talk about space plasma physics, jointly with electric fields measurements and particle measurements . The magnetic field tells us about the wave properties of the plasma, as the electric field also does. At the current time, several kinds of magnetometers are used onboard space plasma physics missions: most often you find a fluxgate to measure DC fields and a searchcoil to measure AC fields. The searchcoil is better than the fluxgate above coarsely 1 Hz. In-situ measurements of plasma wave in Earth environment have been performed since many years by dedicated missions (ESA/CLUSTER (2002), NASA/THEMIS (2007)) and will continue with NASA/MMS mission, a 4 satellites fleet, which will be launched in 2014 with induction magnetometer (see Fig. 1) onboard each spacecraft. Earth is not the only planet with a magnetic field in the solar system. Search coils are for space plasmas physics and how their development still is a challenge.

2. Induction sensor basis

Induction sensor principle derives directly from the Faraday's law:

$$e = -\frac{d\Phi}{dt} \tag{1}$$

where $\Phi = \iint_{(S)} \vec{B}\,\vec{dS}$ is the magnetic flux through a coil over a surface (S).

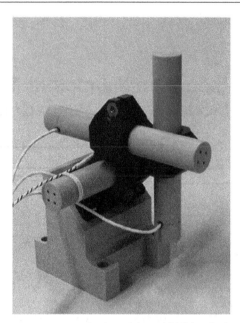

Fig. 1. tri-axis induction magnetometer designed for a NASA mission

The voltage is proportionnal to the time derivative of the flux, thus, by principle, DC magnetic field can not be measured with a static coil. Higher will be the frequency, higher will be the ouput voltage (in the limit of the resonance frequency of the coil).

For (n) coils of section (S), into an homogenous induction magnetic field (B) equation 1 becomes:

$$e = -nS\frac{dB}{dt} \tag{2}$$

This single n turns coils, is designed as air coil induction sensor. As reminded in Tumanski (2007), an increase in sensitivity of air coil induction sensor can be obtained by increasing number of turns (n) or coil surface (S). In applications where the size, mass and performances of the sensor are not too stringent the air-coil induction sensor is an efficient way to get magnetic field variation measurement. Low costs and independancy from temperature variations are two other advantages of the air coil induction sensor.

An important way to improve the sensitivity of an induction sensor consists in using a ferromagnetic core. In that configuration, the ferromagnetic core acts as a magnetic amplifier and coils is wounded around the ferromagnetic core (Fig. 2)

Unfortunately, things are not so easy as ferromagnetic cores have some drawbacks, among them: non-linearity and saturation of the magnetic material, complexity of the design, tricky machining of the ferromagnetic core, whatever the material (ferrite, mu-metal) and the extra costs induced by this part in the design. But the ferromagnetic core still is worth being used as it dramatically increases the sensitivity of the sensor. Thereafter we will consider induction sensors using ferromagnetic cores even if most of our discussion is applicable to the air-coils too.

Fig. 2. Induction sensor using ferromagnetic core.

2.1 Ferromagnetic core

In this chapter, we will start with the description of magnetic amplification provided by a ferromagnetic core. Our description will rely on a simplified modelling of demagnetizing field (Chen et al. (2006); Osborn (1945)). Demagnetizing field energy modelling is still of great importance for micromagnetism studies of magnetic sensors. The search coil demagnetizing field effect study has the advantage to be a pedagogic application and to give magnitude scales to the designer through the apparent permeability concept which is at the source of the magnetic amplification of a ferromagnetic core but which also a way to produce welcome magnetic amplification Popovic et al. (2001). This last point remains a common denominator of many magnetic sensors.

When a magnetic field is applied on a ferromagnetic material, this one becomes magnetised. This magnetization, linked to the magnetic field as expressed by eq. 3, implies an increase of flux density (eq. 4).

$$\overrightarrow{M} = \chi \overrightarrow{H} \tag{3}$$

$$\overrightarrow{B} = \mu_0 \left(\overrightarrow{H} + \overrightarrow{M} \right) \tag{4}$$

The magnetic susceptibility χ can vary from unity up to several tens of thousands for certain ferromagnetic materials. When magnetic field line exit from the ferromagnetic core, a magnetic interaction appears Aharoni (1998) which is opposite to the magnetic field. This interaction, designed as demagnetizing field, is related to the shape of the core through the demagnetizing coefficient tensor $||N||$ and magnetization (cf. eq. 5).

$$\overrightarrow{H_d} = -||N||\overrightarrow{M} \tag{5}$$

By combining, eq. 3, 4 and 5 we can express the ratio between flux density outside of the ferromagnetic body (B_n) and inside (B_{ext}). This ratio, designed as apparent permeability (μ_{app}, cf. eq. 6), depends on the relative permeability of the ferromagnetic core (μ_r) and the demagnetizing field factor ($N_{x,y,z}$) in the considered direction (x, y or z).

$$\mu_{app} = \frac{B_n}{B_{ext}} = \frac{\mu_r}{1 + N_z(\mu_r - 1)} \tag{6}$$

Under the apparent simplicity of the previous formulas is hidden the difficulty to get the demagnetizing field factor. Some empirical formulas and abacus are given for rods in

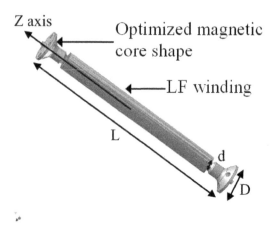

Fig. 3. Ferromagnetic core using flux concentrators.

Bozorth & Chapin (1942) while analytic formulas for general ellipsoids, where demagnetizing coefficient is homogeneous into the volume, are computed in Osborn (1945). However in common shapes of ferromagnetic cores demagnetizing coefficients are inhomogenenous and numerical simulation could be helpfull to guide the design. As explained in Chen et al. (2006) one should distinguish between fluxmetric and magnetometric demagnetizing coefficient. The apparent permeability of concerns should be the one along the coil. To take advantage of the maximum flux, the coil should be localized into 70-90% of the total length depending on the length to diameter ratio of the core Bozorth & Chapin (1942). In the case of rods or cylinders with high aspect ratio (i.e m=length/diameter>>1), the formulas for ellipsoid given in Osborn (1945) allows to get a good estimate of demagnetizing factor in the longitudinal direction (z) :

$$N_z = \frac{1}{m^2} \left(Ln\left(2m\right) - 1 \right) \tag{7}$$

2.1.1 Increase of magnetic amplification using flux concentrator

A way to increase magnetic amplification is to used magnetic concentrators at the ends of the ferromagnetic core (3). Let us consider a ferromagnetic core using magnetic concentrators of length (L), center diameter (d) and ends diameter (D). The classical formula of apparent permeability 6 becomes eq. 8. For a hollow core an extension of this formulas is given in Grosz et al. (2010).

$$\mu_{app} = \frac{B_n}{B_{ext}} = \frac{\mu_r}{1 + N_z \left(L/D \right) \frac{d'^2}{D^2} \left(\mu_r - 1 \right)} \tag{8}$$

For a given set of length, diameter and magnetic material, an increase of magnetic concentrators diameter will lead to a significant increase of apparent permeability (we report increase of apparent permeability higher than 50% in (Coillot et al. (2007)). This increase allows to reduce the number of turns of the winding, all things being equal. Thus the mass of the winding and as a consequence the thermal noise due to the resistance of the winding will be reduced. To take advantage of this improvment, design of the sensor by means of mathematical optimization (Coillot et al. (2007)) is recommended.

2.2 Electrokinetic's representation

Assuming a coil of N turns wounded on single or multi layers and assuming negligible abundance coefficient. The coil exhibits a resistance which can be computed with eq. 9.

$$R = \rho N \frac{\left(d + N\left(d_w + t\right)^2 / L_w\right)}{d_w^2} \tag{9}$$

where ρ is the electrical resistivity, d_w is the wire diameter, t is the thickness of wire insulation, d is the diameter on which coil is wounded and L_w is the length of the coil.

This coil exhibits a self inductance which can be expressed 10 in case of an induction sensor using ferromagnetic core and using Nagaoka formulas in case of an air-coil induction sensor ((n.d.a)).

$$L = \lambda \frac{N^2 \mu_0 \mu_{app} S}{l} \tag{10}$$

where (S) is the ferromagnetic core section, μ_0 is the vacuum permeability and $\lambda = (l/l_w)^{2/5}$ is a correction factor proposed in (Lukoschus (1979)). Designers should notice that practically relative permeability is complex and apparent permeability too, the imaginary part corresponds to loss in the ferromagnetic core. At low field, main mechanisms involved in the imaginary part of the permeability, will be damping of magnetization and domain wall motion in ferrite ferromagnetic coreLebourgeois et al. (1996) while eddy current will be significant in electrically conductive ferromagnetic material like NiFe type. The imaginary part of permeability should be taken into account in the inductance modelling, and also in the following of the modelling detailed in this chapter, since resonance frequency becomes higher than few kHz.

The difference electric potential between each turn of the coil implies an electrostatic field and as a consequence a capacitance because of the stored electric energy between the turns of the coil. For a single layer winding, the capacitance between conductors should be considered. For a multi layer winding the capacitance between layer will be preponderant. In that case and assuming a discontinuous winding (presented on Fig. 5) equation 11 allows to estimate the capacitance of the coil. The capacitance of the continuous winding (see Fig. 5) will not be derived here. Because of his bad performances (explained at the end of this paragraph) we don't recommend its use for induction sensor design. The computation of winding capacitances in continuous and discontinuous windins is detailed pages 254-258 of Ferrieux & Forest (1999) which is unfortunaley in French.

$$C = \frac{\pi \varepsilon_0 \varepsilon_r l_w}{t(n_l - 1)} \left(d + 2n_l(d_w + t)\right) \tag{11}$$

where ε_0 and ε_r are respectively the vacuum permittivity and the relative permittivity (of the wire insulator), (n_l) is the number of layers and the other parameters were defined before.

Now, for a given sensor, the elements of the induction sensor electrokinetic modelling (Fig. 4) can be fully determined.

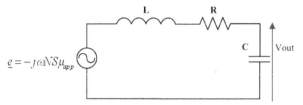

Fig. 4. Représentation électrocinétique du fluxmètre

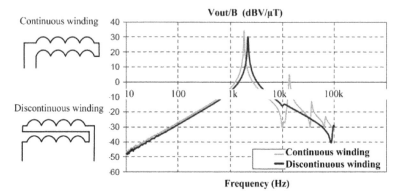

Fig. 5. Induction sensor transfer function for continuous and discontinuous winding (figure extracted from Moutoussamy (2009))

Then, the transfer function between the output of the seameasurable voltage and flux density can be expressed using equation 12.

$$\frac{\underline{V_{out}}}{\underline{B}} = \frac{-j\omega NS\mu_{app}}{(1 - LC\omega^2) + jRC\omega} \tag{12}$$

It shows that induced voltage will increase with frequency until the occurence of the resonance of the induction sensor ($\frac{1}{\sqrt{(LC)}}$). Even if the gain at the resonance can be extremely high it is considered as a drawback since resonance can saturate the output of the electronic conditionner. Moreover the output voltage decreases after the resonance (5). It can be noticed that, in case of continuous winding, multiple resonances appear beyond the main resonance frequency (Seran & Fergeau (2005) & Fig. 5). These multiple resonances do not appear when discontinuous winding strategy Moutoussamy (2009) is implemented. The absence of multiple resonance using discontinuous winding could be due to the homogeneous distribution of the electric field between the layers of the winding.

The methods presented in the next chapter (feedback flux and current amplifier) give an efficient way to suppress the first resonance and to flatten the transfer function of induction sensors.

2.3 Electronic conditioning of induction magnetometers

In this section we will give some details about the classical electronic conditioning associated to induction sensorsTumanski (2007). The feedback flux design and the current amplifier will

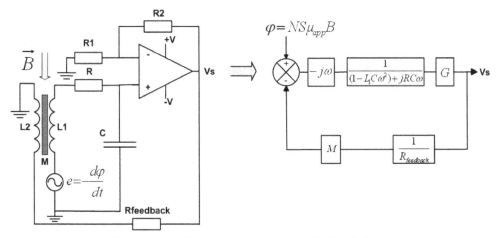

Fig. 6. Principle & bloc-diagram of induction sensor using feedback flux

be presented. They give an efficient way to suppress the first resonance and to flatten the transfer function of induction sensors.

2.3.1 Induction sensor using feedback flux

With a feedback flux added to the induction sensor, as presented on figure 6, the resonance of the induction sensor can be suppressed. A representation of the sensor using bloc diagram makes easy the computation of the transmittance of the feedback flux induction sensor:

$$T(j\omega) = \frac{V_{out}}{B} = \frac{-j\omega NS\mu_{app}G}{(1 - LC\omega^2) + j\omega\left(RC + \dfrac{GM}{R_{fb}}\right)} \tag{13}$$

where M is the mutual inductance between the measurement winding and the feedback winding, R_{fb} is the feedback resistance and $G = (1 + R2/R1)$ is the gain of the amplifier.

Transfer function of induction sensor using feedback flux is illustrated in 8. It demonstrates how the transfer function is flattended.

2.3.2 Induction sensor using current amplifier

Transmittance of the current amplifier (also designed as transimpedance) is expressed according to 14.

$$V_{out} = -R_f I_{in} \tag{14}$$

As current amplifier have a propensation to oscillate a capacitance in parallel is needed to stabilize it (n.d.b). In such case, the transmittance of the current amplifier becomes :

$$V_{out} = -\frac{R_f}{1 + jR_fC_f\omega}I_{in} \tag{15}$$

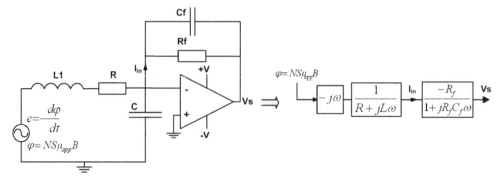

Fig. 7. Principle & bloc-diagram of induction sensor using transimpedance amplifier

In the case of an induction sensor the relation between the induced voltage and the current flowing througth the induction sensor is expressed following:

$$e = (R + jL\omega)I_{in} \tag{16}$$

From the previous equation, is is obvious to obtain a block diagram representation of the induction magnetometer using transimpedance amplifier (7).

Finally, using the bloc-diagram representation 7 the transmittance ($T(j\omega)$) of the induction magnetometer is expressed:

$$\frac{V_{out}}{B} = \frac{jNS\mu_{app}\omega}{R + jL_1\omega} \times \frac{R_f}{1 + jR_fC_f\omega}I_{in} \tag{17}$$

which finally can be written:

$$T(j\omega) = \frac{V_{out}}{B} = \frac{R_f}{R}\frac{j\omega NS\mu_{app}}{1 + \left(\frac{L + R_fC_f}{R}\right)j\omega - \frac{R_f}{R}LC_f\omega^2} \tag{18}$$

The response of the transimpedance amplifier can be either computed using the transmittance equation or computed by simulation software. The comparison of equation 18 with a pspice simulator gives a very good agreement even if the real tansmittance of the operational amplifier modifies sligthly the cut-off frequency. Let us consider an induction sensor designed for VLF measurements, using a 12cm length ferromagnetic core (ferrite with a shape similar to the diabolo juggling prop) and with 2350 turns of 140µm copper wire. The parameters of the electrokinetics can thus be obtained. The parameters of the sensor are summarised in Table 1.

With Rf=470kΩ and Cf=3.3pF for the transimpedance amplifier and Go=100 & Rfb=4,7kΩ for the feedback flux amplifier. We assume the bandwidth of the amplifiers are higher than needed. The transmittance of the transimpedance amplifier exhibits a nice flat bandwidth on 4 decades while the one of the feedback flux is flat on less than 2 decades (Fig. 8).

The comparison between analytic modelling formulaes and pspice simulations validates the analytic formulas proposed respectively for feedback flux induction sensor and transimpedance induction sensor. The analytic formulas are in good agreements with

Characteristics	Value
Ferromagnetic core length	12cm
Ferromagnetic core diameter	4mm
Diabolo ends diameter	12mm
Relative permeability	3500
Turns number	2350
Inductance	0.306H
Resistance	48Ω
Capacitance (incl. 20cm cable)	150pF

Table 1. induction sensor design example

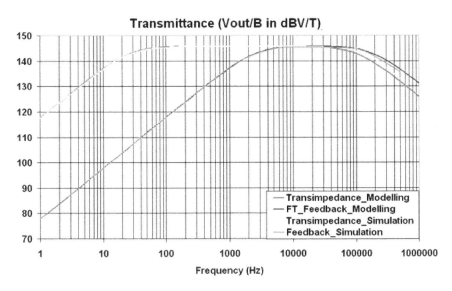

Fig. 8. Transmittance of induction sensor using either a transimpedance or a feedback flux (analytic modelling versus Pspice simulation).

the transmitances reported for the two types of electronic conditionner reported, i.e. the feedback amplifier on one side (Coillot et al. (2010); Seran & Fergeau (2005); Tumanski (2007)) and transimpedande amplifier on the other side (Prance et al. (2000)). The shape of the transmittance can be a reason to have a preference for transimpedance amplifier in some applications especially as the flatness of the transmittance can be increased up to 6 decades by using a compensation network (Prance et al. (2000)), which is simply an integration of the induction signal.

2.4 Noise Equivalent Magnetic Induction (NEMI) of induction magnetometers

Noise equivalent magnetic induction, expressed in $T/\sqrt{Hz}$ is defined as the output noise related to the transfer function of the induction magnetometer 19.

$$NEMI(f) = \sqrt{\frac{PSD_{out}(f)}{T(j\omega)^2}} \tag{19}$$

NEMI is the key parameter for induction magnetometer performances. In most of the design reported the NEMI is computed at a given frequency (often at a low frequency to simplify the problem) while the computation of NEMI on the whole spectrum could permit to adjust the design in case of wide band measurments. The way to compute the NEMI consists in adding different relevant noise sources on the magnetometer schematic. Then, we must consider the transmittance seen by each noise contribution and compute the output PSD. Finally the resulting output PSD will be divided by the transmittance to get the NEMI 19.

For this study we assume that equivalent input noises (both for voltage and current) at positive and negative inputs of the operationnal amplifier are identical and that transmittances of positive and negative inputs signals are closed. Thus, we consider an equivalent voltage input noise (e_{PA}) and an equivalent current input noise (i_{PA}). Since induction sensors are high impedance, amplifier with low input current noise should be preferred, this is the reason why this parameter is often neglected in the modelling. For detailed considerations, use of simulation software is recommended. The purpose of the analytic modelling being to have a friendly tool to help designer.

2.4.1 NEMI of induction magnetometer using feedback flux amplifier

Let us consider first the feedback flux induction magnetometer. Noise sources coming from the sensor and from the preamplifier are reported on the schematic presented on Fig. 9. Using the block diagram representation we can easily get the transmittance "seen" by each noise contribution. Main noise contribution from the induction sensor is the thermal resistance of the winding. Noises sources coming from big volume ferromagnetic core can be neglected at low resonance frequency (Seran & Fergeau (2005)) while Barkhausen noise coming from the magnetic domain displacment can be neglected until reversible magnetization is considered (typ. few mT). We neglect noise contribution from R1//R2 equivalent resistor since this noise can be reduced by design (choosing a small enough R1 resistance).

From the bloc diagram (9), we can easily express the output noise contribution of each noise source, because of the relation between input and output power spectrum density (PSD) through a system characterized by its transmittance ($\underline{T}(j\omega)$):

$$PSD_{out} = \underline{|T(j\omega)^2|}PSD_{in} \tag{20}$$

First, the PSD of the noise coming from the sensor comes essentially from the resistance of the sensor and can be expressed directly as:

$$PSD_R = 4kTR\frac{G^2}{(1-LC\omega^2)^2 + \left(RC\omega + \frac{GM\omega}{R_{fb}}\right)^2} \tag{21}$$

Secondly, the contribution to the PSD of the input voltage noise of the preamplifier (e_{PA}) is:

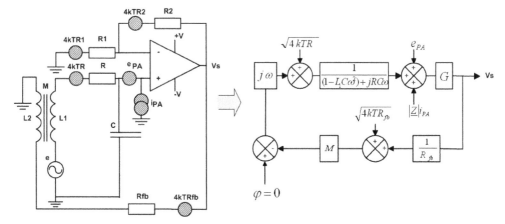

Fig. 9. Feedback flux induction sensor and noise sources.

$$PSD_{e_{PA}} = e_{PA}^2 \frac{G^2\left((1-LC\omega^2)^2 + (RC\omega)^2\right)}{(1-LC\omega^2)^2 + \left(RC\omega + \dfrac{GM\omega}{R_{fb}}\right)^2} \tag{22}$$

Thirdly, the contribution to the PSD of input current noise (converted in noise voltage) of the preamplifier is expressed:

$$PSD_{i_{PA}}(V^2/Hz) = (|Z|i_{PA})^2 \frac{G^2\left((1-LC\omega^2)^2 + (RC\omega)^2\right)}{(1-LC\omega^2)^2 + \left(RC\omega + \dfrac{GM\omega}{R_{fb}}\right)^2} \tag{23}$$

where $|\underline{Z}|$ is the impedance modulus of induction sensor :

$$|\underline{Z}| = \sqrt{\frac{\left(R^2 + (L\omega)^2\right)}{(1-LC\omega^2)^2 + (RC\omega)^2}} \tag{24}$$

By considering the noise contribution of the impedance of the feedback:

$$PSD_{R_f} = 4kTR_f \tag{25}$$

Finally we express the total output noise contribution as a sum of the different contributions:

$$PSD_{out}(f) = PSD_R + PSD_{e_{PA}} + PSD_{i_{PA}} + PSD_{R_f} \tag{26}$$

We can notice that the total ouput noise expressions presented here have only few differences to the one reported in Seran & Fergeau (2005). The objectives of this chapter being double: get the analytic modelling and expose, in a pedagogic way, the method of total noise computation (while equation seems coming from sky in some papers).

By combining 18 with total PSD of noise 31, we can expressed the NEMI defined by 19.

2.4.2 NEMI of induction magnetometer using transimpedance amplifier

Let us now consider the induction sensor with a transimpedance electronic conditionner. The same method as the one presented above is applied to get the NEMI.

In this case, we refer to block diagram of 10. First, the power spectrum densitiy (PSD) of the noise coming from the sensor resistance can be expressed as27

$$PSD_R = 4kTR \times \frac{1}{R^2 + (L_1\omega)^2} \times \frac{\left(R_f\right)^2}{1 + \left(R_f C_f \omega\right)^2} \tag{27}$$

Secondly the contribution to the PSD of the input voltage noise of the preamplifier (e_{PA}) converted in current noise is simply amplified by the transmittance amplifier.

$$PSD_{e_{PA}} = \frac{e_{PA}^2}{R^2 + (L_1\omega)^2} \times \frac{\left(R_f\right)^2}{1 + \left(R_f C_f \omega\right)^2} \tag{28}$$

Thirdly, the contribution to the PSD of the input current noise of the preamplifier is expressed:

$$PSD_{i_{PA}} = i_{PA}^2 \frac{\left(R_f\right)^2}{1 + \left(R_f C_f \omega\right)^2} \tag{29}$$

By considering the noise contribution of the impedance of the transimpedance feedback:

$$PSD_{i_{PA}} = 4kTR_f \frac{1}{1 + \left(R_f C_f \omega\right)^2} \tag{30}$$

Finally we express the total output noise contribution by means of PSD summ:

$$PSD_{out}(f) = PSD_R + PSD_{e_{PA}} + PSD_{i_{PA}} + PSD_{R_f} \tag{31}$$

By combining 18 with total PSD of noise 31, we can expressed the NEMI defined by 19. Finally, the NEMI, which is the first requirement of magnetometers, can easily be determinedd for a given design over the frequency range.

2.4.3 NEMI awards: feedback flux versus transimpedance

Let us consider the example of sensor design presented on 1. We assume the input voltage noise of the preamplifier is about $3nV/\sqrt{Hz}$ while the input current noise is weak $200fA/\sqrt{Hz}$ and that the frequency range is beyond the 1/f noise of the preamplifier. The comparison on Figure 11between the NEMI computed in both cases (transimpedance and feedback flux) exhibit quite close performances in terms of NEMI. Measurment achieved with

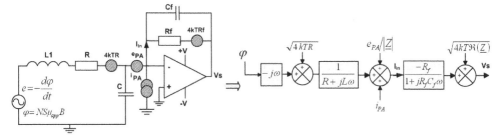

Fig. 10. Induction sensor using transimpedance electronic amplifier: noise sources.

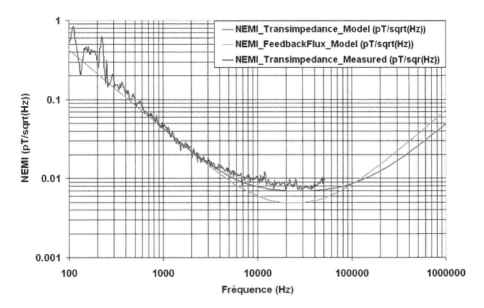

Fig. 11. NEMI comparison: transimpedance amplifier versus feedback flux

a transimpedance preamplifier with same characteristics as the one used in the model and with a sensor identical to the one described in 1 confirm the accuracy of the modelling.

Finally, the NEMI curve can be plotted in both case.

With this design we can get NEMI as low as few fT/sqrt(Hz) using short sensors (12cm in this case). We can notice the minimum of the NEMI around a few 10kHz is (in the design reported here) limited essentially by the input current noise while the NEMI at frequency below 10kHz is limited by the input noise voltage.

We found a good agreement with the NEMI curves of feedback flux induction magnetometer reported in Seran & Fergeau (2005) for an induction magnetometer built for space physics. The increase of NEMI measured above a few 30kHz is the presence of 7th order low-pass filtering. The plots have been superimposed on figure below.

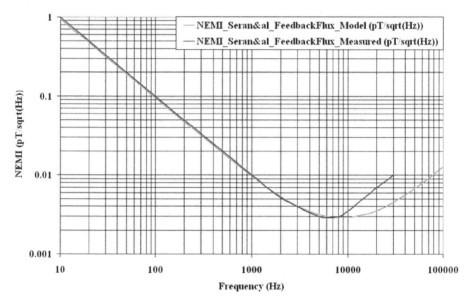

Fig. 12. NEMI of feedback flux amplifier: comparisons between analytic modelling & NEMI reported in Seran & Fergeau (2005)

It would be interesting to compare with other designs, as the one presented in (Grosz et al. (2010); Prance et al. (2000)) but this would require the recollection of datas describing the designs in the mentionned papers.

2.5 Calibration

When designers get their induction magnetometer. Some features of this magnetometer could be useful to know clearly what is measured. First of all, the transfer function and output noise will permit to know the dynamic and the noise equivalent magnetic induction. Next, the directionality of the magnetometer can be a key parameter in applications where the direction of the electromagnetic field must be determined accurately. Another key parameter, which is never mentionned, even if it is of great importance is the sensitivity to electric field. Induction sensors can be sensitive both to magnetic field and electric field and users must care about this last one to avoid to get wrong informations. The electric field sensitivity mechanism is symmetric to the principle of the induction sensor itself. In that case, the electric field will create a current through the wires of the sensor which will be amplify by the amplifier. An electrostatic shielding should surround the sensor and the cable until the preamplifier. This shielding must be refer to a potential (usually the ground), in the same time, this shielding (made of conductive surface) must avoid to allow circulation of eddy current (which would expel the magnetic field at frequency where the skin depth becomes smaller than the thickness of the conductive material). Finally the measurement of the electric field transfer function of the sensor could be an helpfull way to ensure the quality of the electrostatic shielding. As an example, we illustrate, on Fig. 13, a space induction magnetometer where the thermal blanket ensures also electrostatic shielding function.

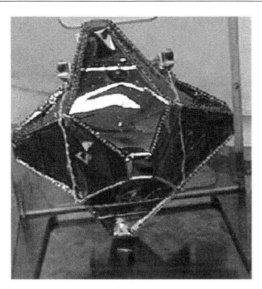

Fig. 13. Tri-axis induction sensor inside a thermal blanket & electrostatic shielding (space experiment CLUSTER).

3. Last and future developments

3.1 Induction sensor bandwidth extension

3.1.1 Dualband search-coil

As mentionned previously, the signal from induction sensor decreases after the resonance frequency while NEMI increases whatever the electronic conditioning is. To bypass this drawback a mutual reducer made of ferromagnetic core is used between two windings Coillot et al. (2010) designed for contiguous frequency bandwidth. It allows to extend the frequency band of measurement using a single sensor. Adjusting a such dual band sensor is not an easy task and its use has sense mainly for applications where mass constraints are stringent.

3.1.2 Cubic search-coil

An interseting way to reduce inductance of induction sensor and thus increase frequency resonance is presented in Dupuis (2005). It consists in implementing induction sensors on the edges of a cube 14.

In such configuration, each axis is constituted in 4 inductions sensors connected in "serie". Let us consider an induction sensor coil requiring a number of turns N. Each induction sensor of the cubic configuration will have $N' = N/4$ turns and, by connecting the single induction sensor in "serie", the total inductance will be proportionnal to $4 * (N')^2$ instead of $(4N')^2$ when the turns are mounted on the same core. Thus, the inductance will be 4 times lower. Since authors claim the capacitance value in the classical configuration and their cubic configuration is the same, the resonance frequency of the cubic induction sensor will be 2 times higher than for classical induction sensor allowing a desirable extend of the frequency range of use.

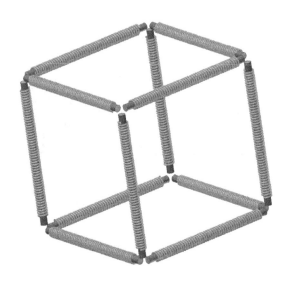

Fig. 14. Cubic induction sensor.

The presence of multiple resoncance beyond the main frequency resonance is not evoked and should be investigated.

The benefit on the apparent permeability value seems also interesting. That comes from the cubic shape which catch more flux. Let us give here a modelling first try.

We first consider cubic induction sensor consituted of ferromagnetic core of length L and diameter d. Because of the cubic shape the demagnetizing coefficient is the same in the 3 directions and we will have:

$$N_x + N_y + N_z = 1 \Rightarrow N_x = N_y = N_z = \frac{1}{3} \tag{32}$$

Then, due the distributed ferromagnetic core on the edges, the flux The total flux caught by the cubic face Φ will be distributed between the four ferromagnetic core with a ratio corresponding to the surface ratio $(L^2/(4d^2))$. If we consider the x direction, we can derive from formula 8 the equation of the apparent permeability:

$$\mu_{app-x} = \frac{\mu_r}{1 + N_x \frac{4d^2}{L^2}(\mu_r - 1)} \tag{33}$$

For high values of relative permeability ($\mu_r \gg 1$ & $N_x \frac{4d^2}{L^2}\mu_r \gg 1$), equation 33becomes:

$$\mu_{app-x} \simeq 3\frac{L^2}{4d^2} \tag{34}$$

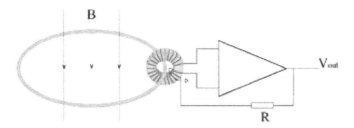

Fig. 15. Magnetic loop antenna description (reprinted with permission from Cavoit (2006). Copyright 2006, American Institute of Physics).

Let's compare the apparent permeability of a cubic induction sensor with the one of a single cylinder ferromagnetic core of 100mm lentgh and 4mm diameter. In the first case, the simplified formula 34 will lead to $\mu_{app-x} = 450$, this apparent permeability is 2 times higher than the one of a single ferromagnetic core.

3.1.3 Magnetic loop antenna

An efficient induction sensor combining a closed loop, in which magnetic field variations generates a current, and a current probe transformer to measure the current flowing trhough the closed loop and feedback this current is presented in Cavoit (2006).

This induction sensor designed for the frequency range from 100kHz up to 50MHz reaches NEMI as low as $0.05 fT/\sqrt{Hz}$ around 2MHz for a 1m*1m square size.

3.2 Miniaturization of induction sensors

3.2.1 Integration of the electronics inside the sensor

A way to reduce size of induction magnetometer is presented in Grosz et al. (2010). It consists in integrating the electronic conditioning inside an hollow ferromagnetic core. They compensate the weak sensitivity of a short ferromagnetic core by using big magnetic concentrators. They try to take advantage of any free volume and they achieve extremely compact and efficient induction magnetometer at the price of a sophisticated mechanical assembly they presented in Grosz et al. (2011).

3.2.2 ASIC Application Specific Integrated Circuit

ASIC in CMOS technology designed for feedback flux induction sensor has been proposed by Rhouni et al. (2010). This miniature electronic circuit (photography of the chip on the left part of Fig. 16), is based on a differential input stage with big size input transistors (MP0 and MP1 on Fig. 16) which allow to reduce strongly the low frequency noise (1/f corner at 10Hz) usually encountered in MOS transistor.

The input noise parameters of this circuit are: $e_n = 4nV/\sqrt{Hz}$ and $i_n < 20fA/\sqrt{Hz}$, close to the best amplifier while the size of the chip is only $2mm \times 2mm$.

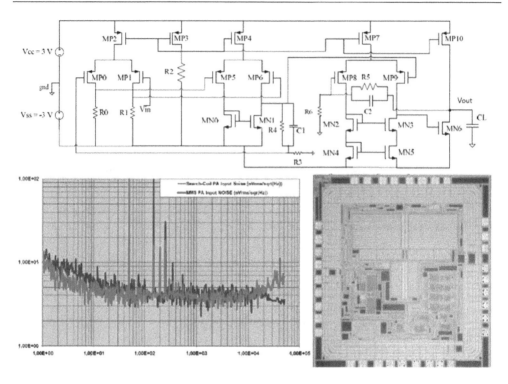

Fig. 16. Design of a feedback flux amplifier ASIC (Fig. on the top : schematic of the ASIC circuit, Fig. on the left side: input noise measurement in $nV/\sqrt{Hz}$, Fig. on the right : photography of the chip)

4. Conclusion

The method presented in this chapter to modelize the induction sensor is based on basic knowledges that can be used to study other types of sensors: elektrokinetics model, noise contributions that must be inventoriated, the computation of the sensitivity... The analytical modelling helps the beginner to become familiar with the sensor and also to manipulate general principles. Even if new technologies can offer excellent performances, in many applications induction sensors remain the best way to achieve AC magnetic field measurements. They will continue to play a role both as induction sensor and combined with other technologies Macedo et al. (2011). One can notice that in some applications, induction sensors have been replaced by other kinds of sensors, like the well known example of Giant MagnetoResistance which have replace coil in the read head of hard disks. Another example is the replacement of the pick-up coil function (like magnetoresistance in squid Pannetier et al. (2004) but also in fluxgates). When the sensitivity does matter the induction sensors still remains an efficient solution. Induction sensors is maybe not the futur of magnetometers but a stone to build this futur.

5. References

(n.d.a).

(n.d.b).

Aharoni, A. (1998). Demagnetizing factors for rectangular ferromagnetic prisms, *Journal of Applied Physics* Vol. 83(6): 3432–3434.

Bozorth, R. & Chapin, D. (1942). Demagnetizing factors of rods, *Journal of Applied Physics* Vol. 13: 320–327.

Cavoit, C. (2006). Magnetic measurements in the range of 0.1-50mhz, *Review of Scientific Instruments* Vol. 77.

Chen, D., Pardo, E. & Sanchez, A. (2006). Fluxmetric and magnetometric demagnetizing factors for cylinders, *Journal of Magnetism and Magnetic Materials* Vol. 306(1): 351–357.

Coillot, C., Moutoussamy, J., Lebourgeois, R., Ruocco, S. & Chanteur, G. (2010). Principle and performance of a dual-band search coil magnetometer : A new instrument to investigate fluctuating magnetic fields in space, *IEEE Sensors Journal* Vol. 10(2): 255–260.

Coillot, C., Moutoussamy, J., Leroy, P., Chanteur, G. & Roux, A. (2007). Improvements of the design of search coil magnetometer for space experiments, *Sensor Letters* Vol. 5: 1–4.

Dupuis, J. (2005). Induction magnetometer, us patent 2005 0156601 a1.

Ferrieux, J.-P. & Forest, F. (1999). *Alimentations a decoupage - convertisseurs a resonance*, DUNOD.

Grosz, A., Paperno, E., Amrus, S. & Zadov, B. (2011). A three-axial search coil magnetometer optimized for small size, low power and low frequencies, *IEEE Sensors Journal* Vol. 11(4): 1088–1094.

Grosz, A., Paperno, E., Amrusi, S. & Liverts, E. (2010). Integration of the electronics and batteries inside the hollow core of a search coi, *Journal of App. Phys.* Vol. 107.

Hayakawa, M. (2007). Monitoring of ulf (ultra-low-frequency) geomagnetic variations associated with earthquake, *Sensors* Vol.7: 1108–1122.

Lebourgeois, R., Fur, C. L., Labeyrie, M. & Ganne, J.-P. (1996). Permeability mechanisms in high frequency polycrystalline ferrites, *Journal of Magnetism and Magnetic Materials* 160(1-3): 329–332.

Lichtenberger, J., Ferencz, C., Bodnar, L., Hamar, D. & Steinbach, P. (2008). Automatic whistler detector and analyzer system, *Journal of Geophysical Research* Vol. 113.

Lukoschus, D. (1979). Optimization theory for induction-coil magnetometers at higher frequencies, *IEEE Transactions on Geoscience electronics* GE-17(3): 56–63.

Macedo, R., Cardoso, F. A., Cardoso, S., Freitas, P. P., Germano, J. & Piedade, M. S. (2011). Self-powered, hybrid antenna-magnetoresistive sensor for magnetic field detection, *App. Phys. Lett.* Vol. 98(10).

Moutoussamy, J. (2009). *Ph. D. Dissertation, Nouvelles solutions de capteurs a effet de magnetoimpedance geante : Principe, Modelisation et Performances.*, Ecole Normale Superieure de Cachan.

Osborn, J. (1945). Demagnetizing factors of the general ellipsoids, *Physical Review* Vol. 7(No. 0): 351–357.

Pannetier, M., Fermon, C., le Goff, G., Simola, J. & Kerr, E. (2004). Femtotesla magnetic field measurement with magnetoresistive sensors, *Science* (304).

Pfaffling, A. (2007). Helicopter electromagnetic sea ice thickness estimation: An induction method in the centimetre scale, *Reports on Polar and Marine Research* 553.

Popovic, R., Randjelovic, Z. & Manic, D. (2001). Integrated hall-effect magnetic sensors, *Sensors and Actuators A* Vol. 91: 46–50.

Prance, R., Clark, T. & Prance, H. (2000). Ultra low noise induction magnetometer for variable temperature operation, *Sensors and Actuators A: Physical* Vol. 85(1-3): 361–364.

Rhouni, A., Sou, G., Leroy, P. & Coillot, C. (2010). A very low 1/f noise asic preamplifier for high sensitivity search-coil magnetometers, *Proceedings of EMSA'10 Conference*, Bodrum (Turkey).

Ripka, P. (2008). Inductive distance sensor for biomedical applications, *Proceedings of IEEE SENSORS Conference*, IEEE, pp. 1230–1232.

Roux, A., Le Contel, O., Robert, P., Coillot, C., Bouabdellah, A., de la Porte, B., Alison, D., Ruocco, S. & Vassal, M. (2008). The search coil magnetometer for themis, *Space Science Review* Vol. 141: 265– 275.

Seran, H.-C. & Fergeau, P. (2005). An optimized low frequency three axis search coil for space research, *Review of Scientific Instruments* Vol. 76: 46–50.

Tumanski, S. (2007). Induction coil sensors - a review, *Meas. Sci. Technol* Vol. 18: R31–R46.

Part 2

Applications

4

Application of Magnetic Sensors to Nano- and Micro-Satellite Attitude Control Systems

Takaya Inamori and Shinichi Nakasuka
The University of Tokyo
Japan

1. Introduction

These days, nano (1 kg – 10 kg) - and micro (10 kg – 100 kg) – satellites, which are smaller than conventional large satellite, provide space access to broader range of satellite developers and attract interests as an application of the space developments because of shorter development period at smaller cost. Several new nano- and micro-satellite missions are proposed with sophisticated objectives such as remote sensing and observation of astronomical objects. In these advanced missions, some nano- and micro-satellites must meet strict attitude requirements for obtaining scientific data or high resolution Earth images. Example of these small satellites are nano remote sensing satellite PRISM, which should be stabilized to 0.7 deg/s (Inamori et al, 2011(a)), and nano astronomy satellite "Nano-JASMINE", which should be stabilized to 1 arcsec (Inamori et al, 2011(b)). Most of these small satellites have the strict constraint of power consumption, space, and weight. Therefore, magnetometers which are lightweight, reliable, and low power consumption sensors are used in the most of these small satellite missions as a sensor for an attitude determination system. In addition, most of these satellites use magnetometers for the attitude control systems with magnetic actuators to calculate required output torque. Furthermore, in some nano- and micro-satellite missions, a magnetic moment is estimated using magnetometers to compensate the magnetic moment and magnetic disturbance. In these small satellite missions, magnetometers play a more important role than conventional large satellites to achieve the attitude control. Although the magnetometers are useful for nano- and micro-satellite attitude control systems, these sensors which are not calibrated are not suitable for an accurate attitude control because of the measurement noises which are caused from magnetized objects and current loops in a satellite. For a precise attitude estimation, many satellites use heavier and higher power consumption sensor such as star truckers and gyro sensors, which are difficult to be assembled to some small satellites. To achieve the accurate attitude control in the small satellite missions, the magnetometers should be calibrated precisely for the accurate attitude control system. This chapter will present what is the requirement for magnetometers in the nano- and micro-satellite missions, how these magnetometers are used in these small satellite missions.

2. Magnetic sensors in nano- and micro-satellite missions

In many nano- and micro-satellite missions, magnetometers are indispensable for attitude control systems and assembled to most of these satellites. In many cases, these small satellites use magnetometers to satisfy the following three requirements: 1) attitude determination using magnetometers, 2) attitude control using magnetic torquers, and 3) estimation of the satellite residual magnetic moment. The magnetometers in most preceding missions featuring large to small sized satellites had the first two requirements. The third requirement is for precise attitude control of the nano- and micro-satellites. In nano- and micro-satellite missions, the magnetic moment, which causes magnetic attitude disturbance, should be compensated to achieve precise attitude control, because the magnitude of magnetic attitude disturbance is lager than the other disturbances in these missions. This chapter presents each requirement of magnetometers in nano- and micro-satellite missions and the methods to achieve these requirements.

2.1 Attitude determination using magnetometers

Magnetometers are generally used for many conventional Low Earth Orbit (LEO) satellite missions for the attitude determination system. For the attitude determination system, the satellites use not only a magnetometer, but also a sun or earth sensor, which can sense the direction of the respective celestial bodies, because satellites cannot get three-axis attitude information with only a three-axis magnetometer. The transformation from a satellite body frame to the reference frame using direction cosine matrix should satisfy the following equation:

$$\mathbf{C}_{bi}\mathbf{V}_k = \mathbf{W}_k \quad (k = 1,.....,n) \tag{1}$$

where V_k is a set of reference unit vectors, which point in the direction of the geomagnetic field, the Sun, and the Earth in the reference coordinate system. These directions are generally calculated with an onboard model such as the International Geomagnetic Reference Field (IGRF) model. W_k is the observation unit vector, which points in the direction observed by the satellite sensors in the body coordinate system. The problem of estimating the attitude from more than two sets of the attitude sensors is referred to as the Wahba's problem (Wahba, 1966). Wahba proposes the following loss function:

$$\mathbf{L}(\mathbf{C}_{bi}) = \frac{1}{2}\sum_{k}^{n} a_i \mid \mathbf{W}_k - \mathbf{C}_{bi}\mathbf{V}_k \mid \tag{2}$$

where a_k are positive values, which define the weight of the sensor information. The problem can be solved with the TRIAD or the QUEST algorithm (Shuster & Oh, 1981) in many satellite missions. The estimated attitude C_{bi} is used for attitude control of a satellite. PRISM and Nano-JASMINE use these methods with a magnetometer and sun sensors for attitude determination. Because it is difficult to remove the effect of the bias noise caused by the residual magnetic moment of the satellite, the satellite generally uses magnetometers for coarse attitude determination.

2.2 Attitude control using magnetic torquers and magnetometers

Magnetic torquers (MTQ) are useful actuators for nano- and micro-satellite attitude control because they are small, lightweight, and low cost actuators with low power consumption.

MTQs are composed of electromagnetic coils, which cause a magnetic dipole moment, interfering with the geomagnetic field, inducing torque for attitude control. The actuator generates torque only in the direction perpendicular to the geomagnetic field. Therefore, the magnetic torquer is not useful for precise attitude control. Most satellites use this actuator for coarse attitude control in an initial phase as a reliable actuator with low power consumption. The actuator is also useful for unloading reaction wheels (Camillo and Markley, 1980).

In this attitude control system using MTQs, the satellites cannot control torque directly, but make use of the magnetic moment to generate torque. Therefore the satellite needs the geomagnetic field information to be able to output the required torque with MTQs. The magnetic moment which should be generated by the MTQs is calculated with measurements performed by magnetometers. A famous attitude control method using a magnetometer is the Bdot method, in which a satellite can stabilize attitude using only a magnetometer in an initial phase when it is difficult to estimate the attitude using sensors after the releasing from a rocket. In accordance with the Bdot method, the magnetic moment of a MTQ is controlled as follows in the PRISM mission.

$$\mathbf{M}_k = \mathbf{K}_p \left(\frac{\mathbf{B}_k - \mathbf{B}_{k-1}}{\Delta t} \right) \tag{3}$$

where K_p is the control gain, B_k is the magnetic field vector obtained by a magnetometer at time step k, and Δt is the step time. Another famous method is the Cross product control method. In this method, an output magnetic moment from a MTQ is computed as follows:

$$\mathbf{M}_k = \mathbf{K}_c \left(\mathbf{T}_c \times \frac{\mathbf{B}_k}{|\mathbf{B}_k|} \right) \tag{4}$$

where K_c is the control gain for the Cross product method, B is the magnetic field vector obtained by a magnetometer, and T_c is the torque requirement calculated by the onboard computer. The method for calculating T_c depends on the objective of the attitude control. For an attitude stabilization with PD controller, T_c can be calculated as follows:

$$\mathbf{T}_c = \mathbf{K}_p (\omega_k - \omega_{ref}) + \mathbf{K}_d \left(\frac{\omega_k - \omega_{k-1}}{\Delta t} \right) \tag{5}$$

where K_p, K_d are control gains, ω_k is an estimated angular velocity at time step k, and ω_{ref} is the target angular velocity. In the Bdot method, the satellite can control angular velocity only to zero and cannot control attitude, but the satellite can stabilize attitude using only a magnetometer. In contrast, using the Cross product method, the satellite can control angular velocity to an arbitrary rate and can also control the direction of the attitude. However, in this case, the satellite requires not only a magnetometer, but also other attitude sensors for the attitude control system. Hence, the Bdot method is useful for attitude stabilization in the phase following ejection from a launcher, when the satellite cannot use many attitude sensors, while the Cross product method is useful for coarse attitude control after this initial phase.

2.3 Residual magnetic moment estimation using magnetometers

Several attitude disturbances are considered for an attitude control system of a nano- or micro-satellite in LEO: magnetic disturbance; air pressure disturbance; solar disturbance; and gravity gradient disturbance. Although, for larger satellites, the effect of the magnetic disturbance is smaller and is therefore not considered in those satellite missions, the magnetic disturbance is larger than the other attitude disturbances for LEO nano- and micro-satellite missions. This is because of the relatively large magnetic moment in satellites with a small moment of inertia and because of the large magnitude of the geomagnetic field in LEO (Inamori et al, 2011(c)). In order to satisfy the strict attitude requirements for LEO nano- and micro-satellite missions, the effect of the magnetic disturbance should be mitigated to achieve a precise attitude control. Because the magnetic disturbance has only a small effect on larger satellites and is not considered in those missions, magnetometers are not used for magnetic compensation. Thus, the magnetic compensation is the new requirement for a magnetometer to achieve precise attitude stabilization for nano- and micro-satellites.

The magnetic disturbance is caused by the interaction of the geomagnetic field and the residual magnetic moment of a satellite. If a satellite estimates the residual magnetic moment and compensates for the magnetic moment with the steady output of a MTQ, the satellite compensates the effects of the magnetic disturbance.

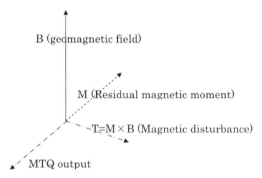

Fig. 1. Overview of the magnetic moment compensation (Inamori et al, 2011(c)).

Fig.1 shows an overview of the magnetic disturbance compensation. In order to compensate for the magnetic disturbance precisely, a satellite must estimate the residual magnetic moment accurately and must have an accurate actuator to generate a magnetic moment. The satellite can estimate a magnetic moment using an extended Kalman filter with magnetometers and gyro sensors. This section presents an estimation technique for the residual magnetic moment using the extended Kalman filter (Inamori et al, 2011(d)). The model of attitude dynamics can be expressed as follows:

$$\mathbf{I}\dot{\omega} = \mathbf{M} \times \mathbf{B} - \omega \times (\mathbf{I}\omega + \mathbf{h}) - \dot{\mathbf{h}} \qquad (6)$$

$$\dot{\mathbf{M}} = diag(-\frac{1}{\tau}, -\frac{1}{\tau}, -\frac{1}{\tau})\mathbf{M} \qquad (7)$$

where I is a tensor of inertia of a satellite, h is the angular momentum of reaction wheels, M is the residual magnetic moment of the satellite, τ is the time constant of the residual magnetic moment, and B is the geomagnetic field vector. The magnetic moment of a satellite is assumed to change over the orbital period, therefore the time constant τ is set to be the orbital period duration. The state vector of the filter can be expressed as follows:

$$x = \begin{pmatrix} \omega \\ M \end{pmatrix} \tag{8}$$

The linearized equations are expressed as follows:

$$\Delta\dot{\omega} = I^{-1}([(I\omega_{ref} + h)\times] - [\omega_{ref}\times]I)\Delta\omega - I^{-1}([B\times]\Delta M) \tag{9}$$

$$\Delta\dot{M} = diag(-\frac{1}{\tau}, -\frac{1}{\tau}, -\frac{1}{\tau})\Delta M \tag{10}$$

which can be arranged in matrix form as

$$\Delta\dot{x} = A\Delta x \tag{11}$$

$$\Delta z = H\Delta x \tag{12}$$

where the state vector is defined as

$$\Delta x = \begin{pmatrix} \Delta\omega \\ \Delta M \end{pmatrix} \tag{13}$$

Matrix A and H can be expressed as follows:

$$A = \begin{pmatrix} I^{-1}([(I\omega_{ref} + h)\times] - [\omega_{ref}\times]I) & -I^{-1}[B\times] \\ 0 & diag(-\frac{1}{\tau}, -\frac{1}{\tau}, -\frac{1}{\tau}) \end{pmatrix} \tag{14}$$

$$H = \begin{pmatrix} E & O \end{pmatrix} \tag{15}$$

where the skew-symmetric matrix $[a\times]$ is defined as follows,

$$[a\times] = \begin{pmatrix} 0 & -a_z & a_y \\ a_z & 0 & -a_x \\ -a_y & a_x & 0 \end{pmatrix} \tag{16}$$

The extended Kalman filter can be computed using the linearized model mentioned above. In the following text, the '-' and '+' superscripts on $\hat{x}$ and P refer to pre-updated and post-updated values, respectively. The over bar indicates an estimate. The time update equation for the state can be expressed as follows:

$$\hat{x}_k^- = f(\hat{x}_{k-1}^-) \tag{17}$$

$$P_k^- = \Phi_{k-1}P_{k-1}\Phi_{k-1}^T + \Gamma_{k-1}Q_{k-1}\Gamma_{k-1}^T \tag{18}$$

where, Φ_k and Γ_k can be expressed as follows:

$$\Phi_k = E + A_k \Delta t \tag{19}$$

$$\Gamma_k = B_k \Delta t \tag{20}$$

The measurement update equations are calculated as follows:

$$K_k = P_k^- H_k^T R_k^{-1} \tag{21}$$

$$\hat{x}_k = \hat{x}_k^- + K_k(z_k - H_k \hat{x}_k^-) \tag{22}$$

$$P_k = P_k^- - P_k^- H_k^T (H_k P_k^- H_k^T + R_k)^{-1} H_k P_k^- \tag{23}$$

The estimated value of residual magnetic moment is applied to the magnetic compensation in orbit. The method is used for PRISM and Nano-JASMINE, which will be presented in section 4.1 and section 4.2.

3. Precise estimation of the geomagnetic field in nano- and micro-satellite missions

As shown in section 2, a magnetometer is used for attitude estimation, attitude control using electromagnetic actuation, magnetic moment estimation in nano- and micro-satellite missions. In these applications of the magnetometers, the effect of the bias error, which is generally caused by surrounding magnetized objects, electrical current loops, and ferromagnetic materials of a satellite, should be cancelled to get accurate magnetic field information. For the precise estimation of the geomagnetic field, calibration of magnetometers is indispensable.

The magnetometer calibration for nano- and micro-satellite missions is not easy on the ground because of two reasons. Firstly, the magnetic sensor should be calibrated with a flight model configuration and flight modes to evaluate the effect of magnetized objects and current loops of devices in a satellite. The calibration of a magnetic sensor assembled in a satellite requires a large facility which has a large magnetic shield to cancel the effect of magnetic field disturbance from devices external to the satellite. Furthermore, the facility should have a device such as a turntable to control the direction between the direction of uniform magnetic field and sensitivity axis of the magnetometer. Such experiments result in a higher cost for the development of the satellite. Secondly, the properties of magnetized objects and current loops of satellite devices can change in orbit after the launch. Some satellite missions reported a change of the residual magnetic moment of the satellites (Sandau et al, 2008, Sakai et al, 2008). This effect can cause a change of the bias error in orbit. For these two reasons, the calibration on ground is not suitable for precise magnetometer calibration. This section will focus on the low-cost, in-orbit calibration for nano- and micro-satellite development.

For in-orbit calibration, several methods have been proposed. These methods can be categorized into the following three groups. The first group is comprised of magnetometer only calibration without attitude sensors. In this method, the norm of the magnetometer measurement vector is compared to the expected norm calculated by geomagnetic filed models. From this comparison, the bias error of a magnetic sensor is estimated (Alonso and Shuster, 2002(a), 2002(b)). In the second group, the predicted magnetic field from geomagnetic filed models is transformed to the body coordinate system with attitude information provided by other attitude sensors (Lerner and Shuater, 1981). In the third group, the methods use several magnetometers to estimate the magnetic moment of a satellite, and then estimate the bias of the magnetometers. In addition, each group has two ways for magnetometer calibration, in-orbit estimation using an onboard computer and offline estimation with telemetry data using a computer on the ground. In offline estimations, magnetometers are calibrated using complicated algorithms on the high performance ground station computers. However, the estimation and compensation of the noise error which have non-steady bias is difficult. This is because the required estimated parameters cannot be updated until the next satellite operation. The effect of time variable bias is caused from current loops of satellite devices and solar batteries. Ferromagnetic materials also cause non-steady bias, because the strength of the bias depends on the direction of the geomagnetic field. These non-steady biases are difficult to estimate using offline estimation. On the other hand, in on-line estimation, time-variable noises can be estimated and compensated, because the parameters for the compensation are updated on-line. In this case, high-spec onboard computers are required for the estimation. With the consideration of the satellite requirements and performance of the onboard computer, satellites must choose the optimal method for magnetometer calibration. This section shows examples with offline estimation using a magnetic model which can be categorized to the first group for the PRISM mission in section 3.1 and on-line estimation using star tracker which is categorized to the second group for the Nano-JASMINE mission in section 3.2.

3.1 Magnetometer only off-line estimation

This subsection presents an example of magnetometer only, offline bias estimation. Because the bias error is as assumed to be a time constant in this method, time variable bias error which is caused from current loops cannot be estimated. The method also cannot compensate for the effect of magnetic anomalies, which is the difference between the Earth magnetic field model and the actual geomagnetic field. Because the method uses the magnetic field model as a reference for the estimation, the magnetic anomaly is observed as bias noise. Although the method is not useful for compensation of the time-variable bias and magnetic anomaly, the method is useful for bias estimation in an initial phase when the satellite cannot use attitude sensors, because the method does not use any other attitude sensors.

In this method, the bias and scale factor errors of magnetometers are estimated using a least squared method. The norm of the IGRF vector should be equal to the norm of magnetometer measurement vector. The difference between the norm of the IGRF model and that of magnetometer shows the bias and the scale factor errors. Using this disagreement information, the bias and the scale factor errors of the magnetometer are estimated by the least squared method on the ground station (Inamori et al, 2011(b)). Cost function J for the least squared method is written as follows,

$$J(\mathbf{s_m}, \mathbf{b_m}) = \sum_{k=0}^{N} \left(\left| \mathbf{IGRF}(x, y, z, t)_{\mathbf{k}} \right| - \left| \mathbf{B}(\mathbf{s_m}, \mathbf{b_m})_{\mathbf{k}} \right| \right)^2 \tag{24}$$

where x, y, and z are the position of a satellite in the reference frame, which can be calculated using two line elements and the SGP4 algorithm. IGRF($\cdot$) is a function for the IGRF model. s_m, b_m are the scale factor and bias of a magnetometer. Magnetometer measurements are modelled as follows,

$$\mathbf{B}_k = s_m \mathbf{V}_{mk} + \mathbf{b}_m \qquad (25)$$

where B, s_m, M_V, and b_m are the magnetometer measurements, scale factors, the output data of an AD converter, and bias noises, respectively. The objective of the method is to estimate the scale factor s_m and the bias noise b_m. These parameters are estimated to minimize the cost function J in equation 24 by the least squared method.

In the PRISM mission, the bias and scale factor errors of a gyro sensor are also estimated using the calibrated magnetometer. The following equations show how to estimate the parameters for the gyro sensor with the calibrated magnetometer. The relationship between the time derivative of the geomagnetic field vector in the body frame and in the reference frame can be expressed as follows:

$$\mathbf{C}_{bi} \frac{d\mathbf{B}_i}{dt} = \frac{d\mathbf{B}_b}{dt} + \omega \times \mathbf{B}_b \qquad (26)$$

where C_{bi} is the direction cosine matrix from a reference frame to a body frame, B_i and B_b are the geomagnetic field vector in the reference frame and the body frame, respectively. B_i changes slowly throughout the orbital period (about 6000 s in LEO). The time derivative of the geomagnetic field in the reference frame is treated as a constant value in a step time in this method. Thus, equation 26 can be written as follows:

$$\frac{d\mathbf{B}_i}{dt} = -\omega \times \mathbf{B}_b \qquad (27)$$

In this equation, the term dB_i / dt should be equal to the term $-\omega \times B_b$. If there are bias and scale factor errors in magnetometer measurements, dB_i / dt and $-\omega \times B_b$ are not equal. In this case, the bias and the scale factor of the gyro are estimated by a least squared method. The cost function J for the gyro calibration is expressed as follows:

$$J(\mathbf{s}_g, \mathbf{b}_g) = \sum_{k=0}^{N} (\frac{dB_{kx}}{dt} - (B_{ky}\omega_{kz} - B_{kz}\omega_{ky}))^2 + \sum_{k=0}^{N} (\frac{dB_{ky}}{dt} - (B_{kz}\omega_{kx} - B_{kx}\omega_{kz}))^2$$
$$+ \sum_{k=0}^{N} (\frac{dB_{kz}}{dt} - (B_{kx}\omega_{ky} - B_{ky}\omega_{kx}))^2 \qquad (28)$$

where ω_k can be expressed as follows,

$$\omega_k = s_g \mathbf{V}_{gk} + \mathbf{b}_g \qquad (29)$$

where s_g, b_g, V are the scale factor, bias noise, and the voltage of the gyro sensors, respectively. These parameters are estimated for calculating minimized the cost function J.

These calibrated magnetometer and gyro sensor measurements are used for attitude estimation and magnetic attitude control system using a MTQ in PRISM mission. The result of the in-orbit calibration in PRISM will be presented in section 4.1.

3.2 Calibration method (on-line estimation with an attitude sensor)

This section presents an example of magnetometer on-line bias estimation using an attitude sensor. For precise attitude control in nano- and micro-satellite missions, a magnetic moment should be estimated and compensated for in order to cancel the effect of the magnetic disturbance. For the accurate magnetic moment estimation, the effect of magnetometer bias should be compensated for. Fig.2. shows simulation results of the magnetic moment estimation of Nano-JAMINE using magnetometer measurement with bias errors. In Nano-JASMINE mission, the magnetic moment should be estimated and compensated for to an accuracy of $1 \times 10^{-4} Am^2$. Fig.2. shows that the bias error should be compensated for to an accuracy of 50 nT. In order to achieve accurate geomagnetic field measurements, some satellites used a magnetometer installed on the tip of a boom in order to remove the effect of the residual magnetic moment of the satellite; however the booms which generally have flexible structure cause attitude disturbances and attitude instability; therefore the booms are not useful for the satellites which aim for precise attitude control. In order to estimate bias error to this accuracy, the effect of time-variable bias and magnetic anomaly should be considered. For this reason, Nano-JASMINE uses on-line bias estimation method using an attitude sensor. Because the method uses the accurate attitude sensor for the calibration, the method is not useful for calibration achieved for attitude determination in the initial phase when accurate attitude sensors are not available.

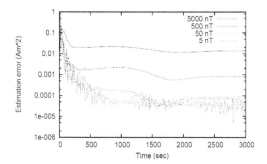

Fig. 2. Simulation result of magnetic moment estimation with magnetometer bias noise.

In this estimation, an Unscented Kalman Filter (Wan et al, 2000) is introduced to estimate the bias noise of magnetometer accurately (Inamori et al, 2010). Firstly, following sets of sigma points are computed.

$$\chi_0 = \hat{x}_{k-1} \tag{30}$$

$$\chi_i = \hat{x}_{k-1} + (\sqrt{(n+\kappa)(P_{k-1}+Q_{k-1})})_i \quad i = 1,\ldots\ldots n \tag{31}$$

$$\chi_i = \hat{x}_{k-1} - (\sqrt{(n+\kappa)(P_{k-1}+Q_{k-1})})_i \quad i = n+1,\ldots\ldots 2n \tag{32}$$

where the matrix Q_k is the process noise covariance, n is the dimension of the state vector, and κ is a scaling parameter. The term $\sqrt{(n+\kappa)(P_{k-1}+Q_{k-1})}$ is calculated using the Cholesky decomposition method. Note that the matrix dimensions are [2n x n]. In this filter, each sigma points calculated in Equation 30-32 are propagated in every filter computation steps. After the propagation of each sigma points, the results are averaged to obtain state vector and covariance matrix accurately in non linear model. The state vector can be expressed as follows in this estimation filter:

$$\mathbf{x} = \begin{pmatrix} \mathbf{q} \\ \mathbf{b} \\ \mathbf{h} \end{pmatrix} \tag{33}$$

where q is quaternion, b is the bias error of magnetometer measurement, and h is the magnetic anomaly of the geomagnetic field. In each step, the state vector and the covariance matrix of the state vector are propagated with dynamics models in an onboard computer. Time update equation is expressed as follows:

$$\chi_{i,k} = \mathbf{f}(\chi_{i,k-1}) \tag{34}$$

$$\hat{\mathbf{x}}_{k-1}^{-} = \sum_{i=0}^{2n} W_i \chi_{i,k} \tag{35}$$

$$P_k^{-} = \sum_{i=0}^{2n} W_i (\chi_{i,k} - \hat{\mathbf{x}}_k^{-})(\chi_{i,k} - \hat{\mathbf{x}}_k^{-})^{\mathrm{T}} \tag{36}$$

where W_i is the weight for the sigma points. The weight W_i can be expressed as follows in this filter:

$$W_0 = \frac{\lambda}{n+\lambda} \tag{37}$$

$$W_i = \frac{\lambda}{2(n+\lambda)} \quad i = n+1,\ldots\ldots2n \tag{38}$$

The function f(x) is obtained by integration using the following differential equations,

$$\dot{\mathbf{q}} = \frac{1}{2} \begin{pmatrix} -[\omega\times] & \omega \\ \omega^{\mathrm{T}} & 0 \end{pmatrix} \mathbf{q} \tag{39}$$

$$\dot{\mathbf{b}} = \mathrm{diag}(-\frac{1}{\tau_b},-\frac{1}{\tau_b},-\frac{1}{\tau_b})\mathbf{b} \tag{40}$$

$$\dot{\mathbf{h}} = \mathrm{diag}(-\frac{1}{\tau_h},-\frac{1}{\tau_h},-\frac{1}{\tau_h})\mathbf{h} \tag{41}$$

where b and h are the magnetometer bias and magnetic anomaly of the geomagnetic field, respectively. The both values, b and h, are assumed to be change in time constant τ_b and τ_h, respectively. After the time update, the state vector is updated with sensor measurements. The predicted observation vector $\hat{y}_{k-1}^-$ and its predicted covariance P_k^{yy} are calculated as follows:

$$Y_{i,k} = h(\chi_{i,k-1}) \tag{42}$$

$$\hat{y}_{k-1}^- = \sum_{i=0}^{2n} W_i Y_{i,k} \tag{43}$$

$$P_k^{yy} = \sum_{i=0}^{2n} W_i (Y_{i,k} - \hat{y}_k^-)(Y_{i,k} - \hat{y}_k^-)^T \tag{44}$$

where h(x) is defined as follows:

$$\hat{y}_k^- = h(\hat{x}_k^-) \tag{45}$$

where y_k is the measurement vector. In this filter, measurements can be obtained as follows:

$$y_k = B_k = C_{bi}(q_k)(H_{igrf} + h_k) + b_k \tag{46}$$

where B_k and H_{igrf} are the geomagnetic field vector calculated with the state vector or obtained by a magnetic sensor, and geomagnetic field vector calculated with the IGRF model, respectively. the C_{bi} is transformation from a satellite body frame to the reference frame, which is calculated from quaternion. The filter gain is computed by the following equation.

$$K_k = P_k^{xy}(P_k^{vv})^{-1} \tag{47}$$

where the innovation covariance is computed by

$$P_k^{vv} = P_k^{yy} + R_k \tag{48}$$

The cross correlation matrix is calculated as follows:

$$P_k^{xy} = \sum_{i=0}^{2n} W_i (\chi_{i,k} - \hat{x}_k^-)(Y_{i,k} - \hat{y}_k^-)^T \tag{49}$$

With the calculated gain in Equation 47, the state vector and covariance matrix are updated with sensor measurement. In this update, the estimated state vector and updated covariance are given by

$$\hat{x}_k = \hat{x}_k^- + K_k(y_k - \hat{y}_k^-) \tag{50}$$

$$\mathbf{P}_k = \mathbf{P}_k^- + \mathbf{K}_k \mathbf{P}_k^{vv} \mathbf{K}_k^T \tag{51}$$

where y_k and $\hat{y}_k^-$ are obtained by a magnetometer measurement and the calculation using the state vector with Equation 46.

These calibrated magnetometer measurements are applied to magnetic moment estimation in the Nano-JASMINE mission. The result of the calibration in Nano-JASMINE will be presented in section 4.2.

4. Example of magnetometer applications for small satellite missions

This section presents the attitude control system of the remote sensing nano-satellite PRISM and micro astronomy satellite Nano-JAMSINE to show examples of the application of magnetometers for nano- and micro-satellite missions. PRISM has three-axis low power consumption magneto-impedance sensor. The magnetometer measurements are used for the attitude estimation and an MTQ is used for attitude control. After completing the main mission, the satellite conducted an in-orbit experiment to demonstrate magnetic moment compensation. Nano-JASMINE has two types of magnetic sensors, a three-axis magneto-resistive magnetometer and a three-axis fluxgate magnetometer. The satellite attitude should be stabilized to within 1 arcsec during observation. In order to satisfy this requirement, the satellite should estimate the magnetic moment and compensate precisely. The fluxgate magnetometer is used for this purpose. The satellite also uses a magnetometer for attitude estimation and magnetic attitude control in initial phase, when the power consumption is limited. A low power consumption magneto-resistive magnetometer is used for this initial phase in this satellite mission. The following subsections show detailed magnetometer application for nano- and micro-satellites with in-orbit data and simulation results.

4.1 Nano remote sensing satellite PRISM

Pico-satellite for Remote-sensing and Innovative Space Missions (PRISM) is a remote sensing nano-satellite, developed at the University of Tokyo. The mission objective of PRISM is to obtain 30 m resolution earth images with a 8.5 kg nano-satellite. The satellite has an extensible boom for an optical system to reduce the total mass of the telescope and to design a compact, light-weight optics system (Fig.3.). Table 1 shows the specification of the PRISM satellite. PRISM has been launched into a Sun-Synchronous LEO as a piggy-back satellite of the GOSAT by a H-IIA rocket in 2009. The satellite attitude should be stabilized to better than 0.7 deg/s during observation. PRISM has gyro sensors and magnetometers for attitude sensors and MTQs as actuators. In the design phase of the project, the attitude was intended to be controlled passively, using only the gravity gradient torque after the boom extension. However, the satellite would have difficulties achieving the attitude requirement because of the magnetic disturbance. To meet the mission requirements, the attitude is determined by sun sensors and magnetometers and controlled actively with MTQs (Inamori et al, 2011(b)).

PRISM has been launched to LEO in January 2009. After extending the boom, PRISM calibrated a magnetometer with the method explained in section 3.1. Fig.4 shows the result

Satellite	Size	192×192×400 [mm³] (Boom folded)
	Total mass	35 [kg]
	Orbit	Sun-Synchronous LEO Altitude 660 [km] Local time 13:00
	Mission	Remote sensing
CDH	CPU	SH7145F (Renesas Technology) H8-3048F (Renesas Technology) PIC-16F877 (Microchip)
Communication	Down	AFSK: 1200 [bps] GMSK: 9600 [bps]
	Up	AFSK: 1200 [bps]
AOCS	Attitude rate requirement	0.7 [deg/s]
	Sensor	Magnetometer (AMI204, three axis) Gyro sensor (tukasa21, three axis) Sun sensor (in-house, two axis)×5
	Actuator	Magnetic torquer (in-house, , three axis)
Optics	Architecture	Refractive
	Diameter	90 [mm]
	Focal length	500 [mm]
	FOV	
	Detector	IBIS-5A, CMOS 1.3 M Pixel

Table 1. Specification of the PRISM satellite.

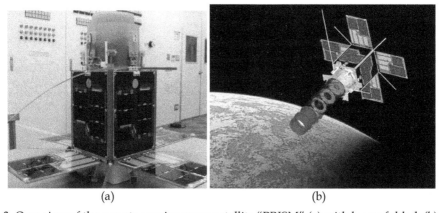

(a)   (b)

Fig. 3. Overview of the remote sensing nano-satellite "PRISM" (a) with boom folded, (b) with boom extended.

of the magnetometer and gyro sensor calibration in orbit. In Fig.4, the magnetometer measurements before the calibration had a bias error. In PRISM project, it was difficult to calibrate assembled magnetometer with a ground facility, thus bias and scale factor were estimated in orbit. After the calibration, magnetometer measurements did not correspond to the IGRF mode in 2000 nT accuracy, because bias of the magnetometer assembled in PRISM had a temperature dependency, which is not considered in the calibration method. After the magnetometer calibration, the bias and scale factor of the gyro sensor was estimated using the least squared method, which is mentioned in section 3.1. Fig.4. shows the angular velocity before and after the sensor calibration. Fig.5. shows the in-orbit result of attitude stabilization before and after the calibration. After the calibration, the satellite attitude was stabilized to an accuracy of 0.3 deg/s, which meets the attitude requirement for the PRISM mission.

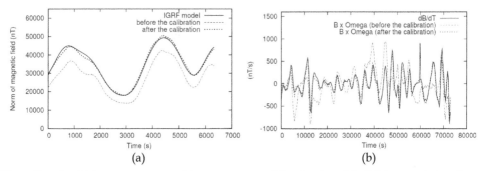

Fig. 4. Result of (a) magnetometer, (b) gyro sensor calibration in-orbit. With permission from ASCE. (Inamori et al, 2011(a)).

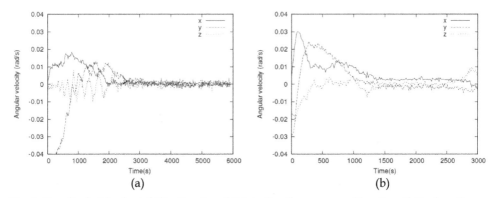

Fig. 5. Result of attitude stabilization in orbit (a) after the sensor calibration, (b) before the sensor calibration.

After the main mission, the satellite conducted an experiment to demonstrate the method to estimate a residual magnetic moment of the satellite. In this experiment, magnetometer and gyro measurement were downloaded to a ground station, then the magnetic moment was estimated using a computer on the ground. After the estimation, the estimated parameters were uploaded to compensate for the magnetic disturbance in orbit. Fig.6. shows the history of angular velocity of a satellite with the magnetic compensation and without the magnetic compensation, respectively. The satellite attitude was stabilized more precisely with the magnetic compensation.

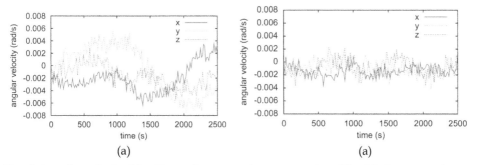

Fig. 6. Angular velocity (a) without the magnetic compensation, (b) with the magnetic compensation (Inamori et al, 2011(d)).

4.2 Micro astronomy satellite Nano-JASMINE

Nano Japan Astrometry Satellite Mission for INfrared Exploration (Nano-JASMINE) is a nano astrometry satellite, developed at the Intelligent Space Systems Laboratory (ISSL), the University of Tokyo cooperation with National Observatory of Japan (NAOJ). The launch is scheduled in 2013. The objective of the Nano-JASMINE mission is to measure the three-dimensional position of stars to an accuracy of 1.8 mas (milliarcsecond), using stellar annual parallax, which is measured by the observed positions of stars at different times of a year. Table 2 shows the specification of the Nano-JASMINE satellite. Nano-JASMINE is outfitted with a telescope with a 5 cm effective diameter and a CCD (Charge Coupled Device Image Sensor) with Time Delay Integration (TDI) which is sensitive to z-band (wavelength ~ 0.9 μm). The satellite measures the position of twenty thousands of stars with a magnitude of larger than 7.5 for all-sky with an accuracy of 1.8 mas. The measurement is used to update the three-dimensional positions and proper motions of the stars in the Hipparcos catalogue. The satellite also demonstrates the space compliancy of the novel type of CCD in which TDI is embedded.

Fig. 7. Overview of the nano astronomy satellite Nano-JASMINE

For the Nano-JASMINE mission, there are two attitude requirements: control the orientation of the telescope and the angular velocity of the satellite. For the requirement of the orientation, the attitude must be controlled to an accuracy of 0.05 degrees. The attitude requirement can be achieved using a star tracker (STT) and a fibre optical gyro (FOG), which have been used in many satellite missions. For the requirement of the angular velocity, the angular rate must be controlled to the Time Delay Integration (TDI) rate in the CCD. The error of the angular velocity with respect to the TDI rate must be smaller than 4×10^{-7} rad/s. This requirement implies attitude sensing and actuation capabilities that are beyond the current state of the art of small satellites. To achieve the mission requirement, the satellite determines the attitude using blurred star images obtained from the mission telescope. Specifically, this method assesses the quality of star images based on how blurred it appears. To get star images using the telescope, the satellite must stabilize the attitude with attitude sensors such as the FOG before using the telescope. Nano-JASMINE stabilizes the attitude using several sensors and actuators before the observation.

Satellite	Size	508×508×512 [mm^3] (When the boom is folded)
	Total mass	35 [kg]
	Orbit	Sun-Synchronous LEO Altitude 750 [km]
	Mission	Infrared Astrometry
CDH	CPU	MicroBraze micro processors×6 configured in Virtex-5×2
Communication	Down	S-band 100 [kbps]
	Up	S-band 1 [kbps]
AOCS	Attitude rate requirement	740 mas/8.8s
	Attitude requirement	0.05 [deg]
	Sensor	Coarse Magnetometer (HMC2003, three axis) Precise Magnetometer (AMI204, three axis) Mems gyro sensor (, three axis) Fiber optic gyro (three axis) Sun sensor (in-house, two axis)×6 Star trucker×2
	Actuator	Magnetic torquer (in-house, three axis) Magnetic canceller (in-house, three axis) Reaction wheel×4
Optics	Architecture	Refractive
	Diameter	90 [mm]
	Focal length	500 [mm]
	FOV	0.5 deg×0.5 deg
	Detector	

Table 2. Specifications of the Nano-JASMINE satellite.

The satellite uses magneto-resistive magnetometer for attitude estimation and an MTQ for actuation in the initial phase after the launch. In this phase, the satellite estimates the attitude using the QUEST method with the magnetometer and sun sensors. After controlling attitude coarsely using the magnetic attitude control system, the satellite calibrates the magnetometer for the estimation of a magnetic moment. Fig.8(a) and Fig.8(b) are simulation results of the bias and magnetic anomaly estimation, respectively. In the simulation, the bias noise is assumed to change throughout an orbital period, because time-variable bias noise has a relationship to temperature and current loops of batteries, which change throughout an orbital period. The magnetic anomaly observed from a satellite is also assumed to change throughout an orbital period, because the magnetic anomaly is fixed with respect to the Earth coordinate system and constant within several days.

In this estimation, both bias and magnetic anomaly errors are estimated with an accuracy of 50 nT. After the estimation of the bias noise, the magnetic moment of Nano-JASMINE is estimated using the extended Kalman filter which is presented in section 2.3. Fig. 9. shows a simulation result of the magnetic moment estimation. In this simulation, the magnetic moment is assumed to change in an orbital period, because the magnitude of current in batteries changes in an orbital period. Based on this estimated value, the satellite compensates for the magnetic disturbance.

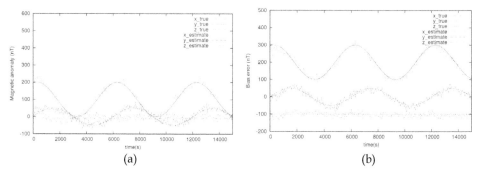

(a)   (b)

Fig. 8. Simulation results of the on-line bias estimation. (a) Estimation of the bias. (b) Estimation of the magnetic anomaly.

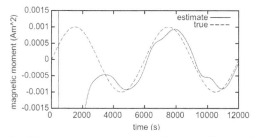

Fig. 9. Simulation result of the magnetic moment estimation (Inamori et al, 2011(c)).

5. Conclusion

This chapter shows the application of magnetometers in nano- and micro-satellite missions. Previously, most of satellites had magnetometers for coarse attitude estimation and attitude control. The magnetometers are also used on nano- and micro-satellite for attitude determination and control system. These days, these small satellites are applied to more sophisticated objectives such as remote-sensing and astronomy missions, which require precise attitude stabilization. For the achievement of the requirements, the satellite must control attitude precisely. Because the small satellites have a small moment of inertia, the magnetic moment which causes magnetic disturbance should be compensated for achievement of accurate attitude control. For the compensation of the magnetic disturbance, the satellites estimate a magnetic moment using a magnetometer. This is the new requirement for magnetometers in the space applications. Magnetometers are indispensable for attitude control and determination as well as magnetic disturbance compensation to achieve the sophisticated objectives in nano- and micro-satellite missions.

6. References

Alonso,R., Shuster, M. D., (2002). Attitude-independent magnetometer-bias determination: a survey, *Journal of the Astronautical sciences*, Vol.50, No.4, pp.453-475, 2002.

Alonso,R., Shuster, M. D., (2002). TWOSTEP: a fast robust algorithm for attitude-independent magnetometer-bias determination, *Journal of Astronautical Sciences*, , Vol.50, No.4, pp.433-451, 2002.

Camillo, P J | Markley, F L (1980). Orbit-averaged behaviour of magnetic control laws for momentum unloading, *Journal of Guidance and Control*, Vol. 3, (November-December 1980), pp. 563-568.

Inamori, T., Sako, N., & Nakasuka, S. (2010). Strategy of magnetometer calibration for nano-satellite missions and in-orbit performance, *AIAA Guidance, Navigation and Control Conference*, (2010), AIAA-2010-7598, Toronto Canada, August.

Inamori, T., Shimizu, K., Mikawa, Y., Tanaka, T., & Nakasuka, S. (2011). Attitude stabilization for the nano remote sensing satellite PRISM, *ASCE's Journal of Aerospace Engineering*, Article in press.

Inamori, T., Sako, N., & Nakasuka, S. (2011). Attitude control system for the nano-astrometry mission "Nano-JASMINE"", *Aircraft Engineering and Aerospace Technology*, (2011),Volume 83, Issues 4, pp.221 – 228.

Inamori, T., Sako, N., & Nakasuka, S. (2011). Compensation of time-variable magnetic moments for a precise attitude control in nano- and micro-satellite missions, *Advances in Space Research*, (2011), Volume 48, Issue 3, pp 432-440.

Inamori, T., Sako, N., & Nakasuka, S. (2011). Magnetic dipole moment estimation and compensation for an accurate attitude control in nano-satellite missions, *Acta Astronautica*, Volume 68, Issues 11-12, (2011), pp2038-2046.

Lerner, G.M., and Shuater, M.D. (1981). In-Flight Magnetometer Calibration and Attitude Determination for Near-Earth Spacecraft, *Journal of Guidance and Control*, Vol. 4, No.5, (September-October 1981), pp. 518-522.

Sakai, S.-i.; Fukushima, Y.; Saito, H. (2008). Design and on-orbit evaluation of magnetic attitude control system for the "REIMEI" microsatellite, Advanced Motion Control, 2008. AMC '08. 10th IEEE International Workshop on , pp.584-589, 26-28 March 2008.

Sandau, R., Röser, H., Valenzuela, A. (2008). Small Satellites for Earth Observation: Selected Contributions, *Springer*, pp185-197, 2008, ISBN: 978-1-4020-6942-0.

Shuster, M. D., and Oh, S. D. (1981). Three-Axis Attitude Determination from Vector Observations, *Journal of Guidance and Control*, Vol. 4, No. 1, (January–February 1981), pp. 70-77

Wahba, G. (1966). A Least Squares Estimate of Satellite Attitude, SIAM Review, Vol. 7, No. 3. (July, 1966), pp. 385-386

Wan, E.A.; Van Der Merwe, R.; , "The unscented Kalman filter for nonlinear estimation," *Adaptive Systems for Signal Processing, Communications, and Control Symposium 2000. AS-SPCC. The IEEE 2000* , vol., no., pp.153-158, 2000

5

The Application of Magnetic Sensors in Self-Contained Local Positioning

Chengliang Huang and Xiao-Ping Zhang
Ryerson University
Canada

1. Introduction

Nowadays, GPS (Global Positioning System) has been widely used to position airplanes, ships, automobiles, migrating animals, and human beings, etc. However, GPS has its own limitations. Other than the user groups for whom GPS is operated, there is no service level agreement with anyone else. As more and more people become dependant on GPS in more and more countries, the political control of GPS raised doubt about its value. Moreover, in the event of system failure, equipment depending on GPS, such as airplanes or ocean-going vessels, needs alternative positioning systems. Furthermore, GPS relies on complicated infrastructure and cannot provide sufficient accuracy, stability and coverage for environments such as indoors and highly urbanized areas (known as "urban canyon") due to the lack of line-of-sight (LOS) to the satellites of GPS. As a result, most commercial and social activities taking place in such environment can not benefit from positioning services offered by GPS.

Local positioning systems (LPS) are intended to provide a back-up for applications where GPS does not perform sufficiently to the requirements of the users. As indicated by the name, LPS can only provide positioning services within a limited area. Unlike GPS, LPS provides positioning services in communities, campuses, urbanized areas and inside buildings. LPS can be categorized into two groups: indoor positioning systems and outdoor positioning systems. The former provides positioning services inside buildings while the latter provides the services outdoors.

LPS can provide positioning services in a wide range of applications, including personal guidance, people-finding, mobile information, moving object management and so on. LPS is also essential for augmented reality and ubiquitous computation. Typical applications of LPS are as follows personal navigation assistances (PNA), personal guidance (PG), LPS tags, people-rescue and intelligent transportation services (ITS).

PNA is a portable electronic device with both positioning function and navigation capability. A PNA based system is more flexible than GPS in providing optimal route descriptions for pedestrians because they only rely on the same maps for both in-vehicle and on-foot scenarios. On the other hand, as an interactive portable device, PG can help people with effective routing, or real-time reservation, and journey updating to inform the concerned parties of the latest situation on the journey. With this service, traveling will be easier and safer for people.

LPS can find the location of people equipped with an LPS tag, such as business men traveling in urban areas, workers working on a construction site, fire fighters in operation, such that the pertinent activities can actively monitored and coordinated for better effectiveness, efficiency and safety. An important application of LPS in public services is people-rescue, such as urban search and rescue (USAR). When disasters such as earthquakes, cyclones, tornadoes and floods occur, some specialized organizations, governments and private companies will dispatch task forces immediately. The task force will firstly locate and extricate victims entrapped, then provide and conduct first aid care. Those victims may be entrapped in confined spaces such as collapsed structures and trenches, mines and transportation accidents. Obviously, the positioning of the victims is critical for an efficient rescue action.

A typical application of LPS in transportation is so-called ITS. This new technology is still being developed to control traffic jam and improve road safety. ITS also reduces vehicle wear, travel time, and fuel consumption through managing competing factors such as vehicle type, load, and routes. Other than conventional services such as electronic toll collection and emergency vehicle notification, sophisticated services are also provided to control urban traffic and to guide drivers with automatic routes for optimal performance. LPS, integrated with GPS, makes these services possible.

In the long run, both the technology and society environments will be changed greatly. Integrated circuits (IC) will be faster, smaller and cheaper according to Moore's law. The positioning will be more accurate and robust with advanced algorithms and new sensors. Building automation and indoor three-dimensional (3D) dynamic mapping will illustrate and control the whole building environment in much greater details such that both the quality of indoor environment and the operation cost-effectiveness will be significantly improved. Virtual world and real world will be merged to create new sports, games, entertainment and arts. People will have more wealth, enjoying more social networking, and receive more cares. As a result, LPS will surely have more and more applications. Personal positioning devices will be popular to every working persons and school students. Value added services will be provided through LBS (Location Based Services) and ITS. GIS (geographic information system) and LPS will be combined to provide quick emergency response and automatic rescuing. Much smarter supply chain management, traffic management, and work force management will be available to businesses. LPS may induce a commercial exploitation in the future.

2. Strap-down inertial positioning

Some LPS systems reply on certain infrastructures and are difficult to setup and maintain. They are not suitable for applications that require quick setup, such as fire fighting operations, and emergency rescue services. We recommend self-contained LPS to locate objects residing or moving within a covered area. These objects are equipped with intelligent sensor modules, normally in form of microelectromechanical system (MEMS) component that can sense both the direction and speed of movement and compute the position of the objects within a given coordinate system. In this section we introduce a self-contained pedestrian tracking technique for the establishment of a complete infrastructure-free positioning system. The technique is the synergism of two existing technologies of dead reckoning and inertial positioning. In the implementation of such a technology, the magnetic sensors take an important role in achieving acceptable performance.

2.1 Dead reckoning

Examples of existing self-contained LPS systems include dead reckoning systems for marine navigation (McBrewster, 2009), wheel sensing systems for automobiles (Thiessen & Dales, 1983), and pedometers for human beings (Chiang, 2007). Dead reckoning is a method of estimating an object's present position by projecting its courses steered and speeds over ground from the last known position. The dead reckoning position is only a rough approximation because it does not consider the effect of leeway, currents, helmsman error, or gyro error.

All dead reckoning systems compute the position of an object based on the last known position and the relative movement from that position. The relative movement is calculated according to the direction and speed of motion. This principle is employed in the aforementioned three existing self-contained LPS. Despite the advantage of independence on complicated infrastructure, these systems normally suffer from low accuracy.

2.2 Inertial positioning

An inertial positioning system (IPS) indirectly obtains the speed and direction, or displacement, through integrating the measured accelerations and the angular velocity of an object over time. Accelerometers and gyroscopes are used to continuously measure and record translational and rotational motions. Thus inertial positioning and dead reckoning are different in the data acquired (acceleration v. s. speed, angular velocity v. s. course direction) and the devices used to measure these data (e. g. accelerometer v. s. speedometer). For a dead reckoning system, the distance can also be acquired through direct estimation. For example, in a pedometer, the step length is simply estimated, not measured or calculated. The distance is estimated by multiplying the estimated step length and the number of steps counted. Generally speaking, inertial positioning, or inertial navigation, is a modern technology superseding the dead reckoning, a relatively older technology.

A traditional inertial positioning system is gimbaled (Sarton and George, 1959). A mechanical device called gimbal-stabilized platform is used to establish a reference system in vehicles such as submarines, surface vehicles, aircrafts and space crafts. The sensors, the gyroscopes and accelerometers, or pick-ups, are mounted up on the stabilized platform to sense specific forces. The advantage of gimbaled systems is that the computation of position, velocity, orientation and angular velocity is less complex.

As shown in Fig. 1, with the initial local velocity $\mathbf{v}^1(0)$ and initial local displacement $\mathbf{s}^1(0)$ of an object known respectively, we are able to track the position of the object by subtracting the acceleration due to gravity, $\mathbf{g}$, and then integrating the remaining acceleration $\mathbf{a}^1(t)$-$\mathbf{g}$, once over a time period t to obtain velocity, and twice to obtain displacement in the local geographic frame (l-frame):

$$\mathbf{v}^1(t) = \mathbf{v}^1(0) + \int_0^t \left(\mathbf{s}^1(\tau) - \mathbf{g} \right) d\tau \tag{1}$$

$$\mathbf{s}^1(t) = \mathbf{s}^1(0) + \int_0^t v^1(\tau) d\tau \tag{2}$$

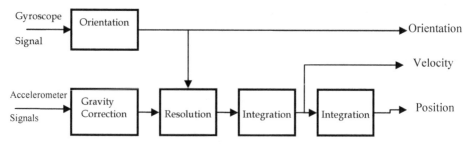

Fig. 1. Traditional inertial positioning algorithm

Using the rectangular rule, the above integration can be implemented in discrete-time form , denoting δt as the integration interval:

$$\mathbf{v}^1(t + \delta t) = \mathbf{v}^1(t) + \left(\mathbf{a}^1(t + \delta t) - \mathbf{g}^1\right) \cdot \delta t \tag{3}$$

$$\mathbf{s}^1(t + \delta t) = \mathbf{s}^1(t) + \mathbf{v}^1(t + \delta t) \cdot \delta t \tag{4}$$

Unfortunately, there are several problems with the gimbaled systems. Other than frictions from bearings and the dead zones of motors, extra power is needed to align the platform with the navigational frame. Moreover, gimbaled systems need high quality electro-mechanical parts including motors, slip rings and bearing, recalibration which is difficult, and regular maintenance by certified personnel in a clean room through a lengthy recertification process. Consequently, the traditional position systems are mostly used for airplanes, vessels, and intercontinental ballistic missiles.

2.3 Strap-down inertial positioning

The strap-down IPS replaces the traditional gimbaled system for low-cost applications. The accelerometers and rate gyroscopes are rigidly mounted in the body of a tagged object thus there is no relative movement between them. This is a major hardware simplification compared with the stabilized platform in tradition system. The most significant advantage of the strap-down IPS in comparison with the gimbaled IPS is the considerably reduced size and weight, which normally result in lower cost, power consumption, and hardware complexity.

However, the processing of inertial navigation is more sophisticated. Fundamentally, for IPS, a number of Cartesian coordinate reference frames, such as the body frame (b-frame), positioning frame and local geographic frame (l-frame) have to be rigorously defined and precisely related. We define the b-frame as an orthogonal axis set, which is aligned with the roll, pitch and yaw axes of the body to be positioned. We also define the positioning frame as an l-frame which has its origin at the original position for positioning and axes aligned with the directions of north, east and the local vertical (down). In a strap-down IPS, the computational complexity is increased because output data are measured in the b-frame rather than the l-frame. An algorithm has to be applied to keep track of the orientation of the sensor module (and the object) and rotate the measurements from the b-frame to the l-frame. Fig. 2 shows this procedure.

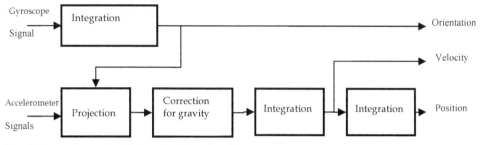

Fig. 2. Strap-down inertial navigation algorithm

2.3.1 Reference frames and rotations

Denoting axes x^l, y^l, and z^l respectively represent north, east and down in the l-frame fixed to the surface of the earth, the orientation of the device will be estimated in this coordinate frame. Axes x^b, y^b and z^b represent the orthogonal axes in the b-frame.

In order to describe the components of a vector of arbitrary orientation in l-frame with respect to the original b-frame, or, in a new frame with respect to an old frame, after a certain rotation, a few rotation representations must be applied. We use the direct cosines representation to develop algorithms for tracking the orientations of sensors. To do this, a direct cosine matrix (DCM) has to be established by transformation of three sequential rotations from the original axes in the l-frame to the new body axes. These rotations continue as a Ψ rotation about z; a θ rotation about y, resulting from the first rotation; and, finally, a ϕ rotation about x, resulting from the second rotation. The resulting DCM is as follows:

$$\mathbf{R} = \begin{pmatrix} \cos\theta\cos\psi & -\cos\phi\sin\psi + \sin\phi\sin\theta\cos\psi & \sin\phi\sin\psi + \cos\phi\sin\theta\cos\psi \\ \cos\theta\sin\psi & \cos\phi\cos\psi + \sin\phi\sin\theta\sin\psi & -\sin\phi\cos\psi + \cos\phi\sin\theta\sin\psi \\ -\sin\theta & \sin\phi\cos\theta & \cos\phi\cos\theta \end{pmatrix} \quad (5)$$

2.3.2 Tracking the orientation of a moving object

The orientation of a moving object is tracked by integrating the angular velocity signal obtained from rate gyroscope in the swing phase of walking.

At a time instant t, let $\omega = [\omega_x\ \omega_y\ \omega_z]^T$ be the corresponding angular velocity sample, denote the norm of angular velocity as ξ, i.e., $\xi = \|\omega_x, \omega_y, \omega_z\|$, and a matrix

$$\Omega = \begin{pmatrix} 0 & -\omega_z & \omega_y \\ \omega_z & 0 & -\omega_x \\ -\omega_y & \omega_x & 0 \end{pmatrix} \quad (6)$$

With a short sampling interval δt, the DCM at time t, $\mathbf{R}(t)$, can be calculated from the DCM at time $t - \delta t$, $\mathbf{R}(t - \delta t)$, according to below equation

$$\mathbf{R}(t) = \mathbf{R}(t - \delta t) \cdot \left(\mathbf{I} + \frac{\sin(\xi\delta t)}{\xi}\Omega + \frac{1 - \cos(\xi\delta t)}{\xi^2}\Omega^2 \right) \quad (7)$$

With the DCM updated, it becomes possible to project the acceleration signal $a^b(t)$ from the accelerometers in the b-frame into acceleration $a^l(t)$ in the l-frame:

$$a^1(t) = R(t) \cdot a^b(t) \tag{8}$$

After the local acceleration is obtained, the velocity and the displacement of the object can be determined according to (1) and (2) respectively to locate the object.

3. Inertial sensors

As mentioned above, inertial positioning relies on the measurement of accelerations and angular velocities. The estimate of changes in rotated angles can be obtained through integrating the angular velocity from a gyroscope over time. The estimate of changes in velocities and positions in the l-frame can be obtained through integrating the local acceleration over time once and twice respectively. The local acceleration is the projection of body acceleration, measured from accelerometers, onto the l-frame.

3.1 Gyroscopes and accelerometers

A wide range of sensors can detect or measure angular movement. These devices range from the conventional mechanical gyroscopes and optical gyroscopes, to the ones based on atomic spin.

An accelerometer measures the physical acceleration it experiences relative to freefall, not to the coordinate systems. There are many types of accelerometers. Generally, all these devices are amenable for strap-down applications. However, the accuracy ranges widely from micro-g to fractions of g due to the variety of designs. Acceleration can be internally measured without using any external references, which are needed for the measurement of velocity. This is why it is preferable to develop algorithms based on acceleration for LPS.

MEMS technology, used to form structures with dimensions in the micrometer scale, is now being employed in manufacturing state-of-the-art MEMS-based inertial sensors.

3.2 Performance evaluation

Practically, the measuring errors of accelerometer and rate gyroscope result in errors during positioning. Assuming the average output from an accelerometer when it is not undergoing any movement is b_a, the accumulated positioning error after double integration over time period t is $e_s(t)= b_a t^2/2$. This means that the positioning error caused by a constant bias of accelerometer increases quadratically through double integration over time. Such an error is called integral drift.

Similarly, assuming the offset of the output angular velocity from true value is b_g, the error of a gyroscope over time period t, we have $e_g(t)= b_g t$. This linearly increasing angular error, i.e. orientation error, can cause the error of the rotation matrix $R(t)$. The resulting incorrect projection of acceleration signals onto the global axes causes two problems. Firstly, the acceleration is integrated in the incorrect direction, causing interaction of velocity and position between (not along) axes. Secondly, acceleration due to gravity can not be removed completely. The residue acceleration originated from gravity will become a "bias" to the true acceleration due to movement of the object.

In the strap-down navigation algorithm, the acceleration due to gravity, **g**, is deducted from the globally vertical acceleration signal before integration. When there are angular errors, there are tilt errors. Here a tilt error refers to the angle between the estimated vertical direction and true vertical direction. The tilt error e_ϕ in radian will cause the projection of gravity onto the horizontal axes, resulting in a component of the acceleration due to gravity with magnitude: $e_a{}^h=g\times\sin e_\phi$. This component can be treated as a residual bias due to gravity, remaining in the globally horizontal acceleration signals. In the mean time, in globally vertical axis, there is a residual bias of magnitude: $e_a{}^v=g\times(1-\cos e_\phi)$. Fortunately this problem is much less severe because for small e_ϕ, we have $e_a{}^v\approx0$. Therefore, a small tilt error will mainly cause positioning error in the globally horizontal plane.

In some cases, Such as human walk, the mean absolute acceleration measured is much smaller than the magnitude of gravity. As a contrast, a tilt error of 0.05° can cause a component of the acceleration due to gravity with magnitude near g/1000. This residue bias can cause a positioning error of 15.4 meters after only a minute of integration, or error of 0.49 meter after only 10 seconds (as demonstrated by the red dashed line in Fig. 3). Therefore, gyroscope errors, which propagate in the positioning algorithm, are critical errors affecting the accuracy of pedestrian tracking. Before the development of the algorithm presented in this chapter, it was believed that positioning with data from inertial sensors was not possible due to the quadratic growth of errors caused by sensor drift during double integration.

4. Integration of PDR with inertial positioning

Precise tracking of people, especially of first responders, in a harsh environment, remains an open research area. For a practical person tracking system, the measuring device of the system should be portable or wearable for the person being tracked. Inertial positioning can be applied in people tracking theoretically. However, the integral drift of accelerometers and gyroscopes, as well as the tilt errors of gyroscopes, make the inertial positioning not practical. As a result, as mentioned earlier, the characteristic of the movement of the person should be incorporated to improve positioning accuracy. If pedestrian dead reckoning (PDR) is integrated with IPS, the performance will be significantly improved, as shown is Fig. 3.

4.1 Pedestrian dead reckoning

PDR is the application of dead reckoning in pedestrian tracking. Due to the difficulty in measuring the speed of a walking person directly, PDR estimate a pedestrian's present position based on estimated step length and heading from the last known position. Using probability models, Mezentsev et al (2005) confirm that the main source of errors in a PDR system is related to the estimation of the step length and the heading. Different random noise models can be applied to heading estimation and step length estimation (Mezentsev et al, 2005). Suppose the modeled mean step length s is a constant, and the ith length error ω_i is normally distributed with a standard deviation σ, then the true ith step length satisfy $s_i=s+\omega_i$. Assuming the step errors are uncorrelated during a walk, the distance error variance after N steps is

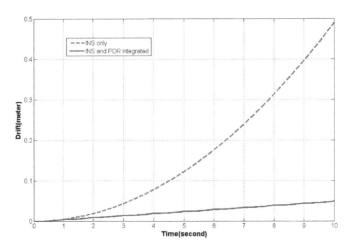

Fig. 3. Performance of positioning with IPS and PDR integrated

$$\sigma_s^2 = Var\left(\sum_{k=1}^{n}\omega_k\right) = \sum_{k=1}^{n}E(\omega_k^2) = N\sigma^2 \tag{9}$$

Hence, the error variation equals the product of the step count and the variation of the driving noise. Accordingly, the position error maximizes when the pedestrian walk continuously along a straight line with a constant step length error. Therefore, if not assisted by other methods, PDR is not able to provide precise positioning. For further improvement of accuracy, the kinetics of human walk is investigated.

4.2 Human gait and phase detection

According to Rose and Gamble (1994), the human motion during a walk is cyclical and repeatable. Such a basic pattern is significantly constant among individuals. The gait cycle during human walk comprises two phases: a stance phase and a swing phase. In the stance phase, a foot keeps in touch with the ground from the time when the heel strikes the ground to the time when the toes leave the ground. The swing phase starts when the foot is taken away and carried forward. When the foot touches the ground again, the swing phase ends and the next stance phase starts. Each gait cycle is completed and restart in such a way. The other foot performs the same process, but lags 180° behind. Fig. 4 shows how the stance phase is started and ended. For a certain point at the front part of sole, its ground speed is almost zero from T_3 to T_5 (the period ΔT). If this time period ΔT can be dependably detected every stride, referential data may be acquired and used to reset computations and get rid of accumulating errors, the critical problem in pedestrian tracking.

Huang et al (2010) advocate the integration of inertial navigation and pedestrian dead reckoning. With the stance phase detected, the velocity and angles can be updated in the stance phase and swing phase respectively. Thus both the integral drift of velocity and positions, as well as the tilt error caused by integral errors of rotated angles, can be greatly reduced. The positioning performance of the integration is illustrated in Fig. 3. The details of these updates are explained in the sub-sections that follows.

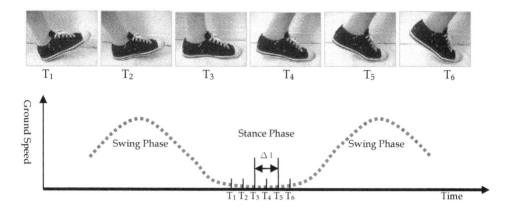

Fig. 4. Stance phase in a stride

In order to detect stance phase we can use sensor data, including all three components of angular velocity, acceleration, and even magnetic field, provided by a typical inertial/magnetic sensor module. Because the y-axis angular rate is the most significant indicator of stride events, Huang et al (2010) use the y-axis angular rate to detect the stance phases instead.

4.3 Zero-Velocity update and swing phase integral error reduction

Theoretically, the velocity in the l-frame is zero in the stance phase. In practice, zero-velocity updates (ZUPT) are performed when the acceleration and rate gyro measurements drop below empirically determined thresholds for a certain period of time. The main purpose of these updates is to mitigate both movement measured in the stance phase of a gait and increasing integral drifts. ZUPT is used by Stiring and Fyfy (2005), Mezentsev and Lachapelle (2005). Stirling and Fyfy stopped integrating and reset the velocity before each swing phase. The positioning accuracy of their system is of 10 to 20 percent of distance traveled.

In order to obtain a higher tracking accuracy, Huang et al (2010) further use additional methods to remove integral drifts. Some methods should be applied in the swing phase of each stride.

4.4 Orientation update using magnetic sensors

Huang et al (2010) update orientation using magnetic field signals from magnetic sensors, i.e., magnetometers.

In order to determine the orientation of the sensor module for rotation representation and transformation between the l-frame and the b-frame, the Euler angles must be determined. In the swing phases, the Euler angles are calculated through integration of the angular velocity over time assuming that the original orientation of the sensor module is known. However, the original orientation is usually unknown in reality. In addition, the gyroscope signal summed during integration is perturbed by gyroscope bias. Thus the accuracy of the orientation will decay overtime. Therefore, Huang et al (2010) use magnetometers to detect the magnetic field of the earth. Using data from magnetometers, together with gyroscopes and accelerometers, the orientation in the beginning of each human walk can be found. Similarly, the orientation of the sensor module in stance phase, when the sensor module is in a near-static status, can be updated in each gait. With the updates of orientation in stance phases, the integral drifts of rotated angles, which increase linearly over time, are greatly reduced.

To find the near-static orientation, the Euler angles ψ, θ and φ must be calculated. Because gyroscopes cannot be used in a near-static status, signals from accelerometers and magnetometers have to be used to calculate the orientations of the sensor module. However, these angles cannot be obtained directly. To resolve this problem, the tilts and heading of the sensor module need to be investigated.

Both x-tilt α and y-tilt β of the sensor module, or the sensor on it, as shown in Fig. 5, can be found simultaneously using acceleration outputs of all three axes as below:

$$\alpha = \text{atan} 2\left(a_x, \sqrt{a_y^2 + a_z^2}\right) \tag{10}$$

$$\beta = \text{atan} 2\left(a_y, \sqrt{a_x^2 + a_z^2}\right) \tag{11}$$

where the two-argument function, atan2, is a variation of the arctangent function. For any real arguments x and y, which are not both equal to zero, atan2(y, x) is the angle in radians between the positive x-axis of a plane and the point given by the coordinates (x, y) on it.

The heading of a pedestrian, thus that of the sensor module on him/her, can be calculated using the x-axis and y-axis magnetic field outputs if the sensor module is level in a horizontal plane. As shown in Fig. 6, only the x and y components of the Earth's magnetic field, are used when determining the heading of sensor module. The heading γ is defined by:

$$\gamma = \text{atan} 2(h_y, h_x) \tag{12}$$

where h_x, h_y and h_z are components of magnetic field on x, y, z axis respectively.

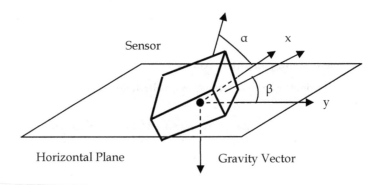

Fig. 5. x-tilt and y-tilt assignments relative to ground

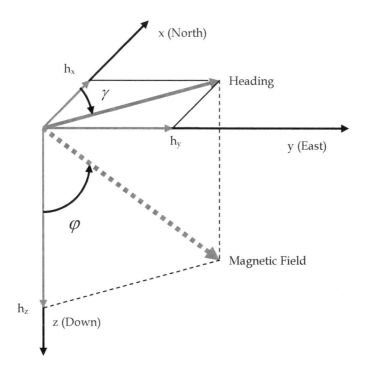

Fig. 6. Heading defined in the horizontal x-y plane

No matter whether a person is standing or walking, the sensor module attached to the human body is not confined to a flat and level plane. This makes it more difficult to determine the heading direction, because the tilt angles of the x-axis and the y-axis are always changing and the sensor module cannot stay horizontal to the earth's surface. Errors caused by tilt angles can be large if not compensated.

To compensate the tilt of the sensor module when it is in an arbitrary position, the l-frame is transformed into the b-frame through three rotations with Euler angles ψ, θ and ϕ in sequence. The DCMs of the rotations are denoted as $R_1(\psi, z)$, $R_2(\theta, y)$ and $R_3(\phi, x)$ respectively. These Euler angles are related to the tilt but they are not all the tilts of the sensor module. The algorithm (Huang et al 2010) is as follows.

Rotations of the gravity vector from the l-frame to the b-frame are given by

$$g^b = R^T \cdot g^l = \begin{bmatrix} -\sin\theta & \sin\phi\cos\theta & \cos\phi\cos\theta \end{bmatrix}^T \tag{13}$$

where $g^b = (g_x\ g_y\ g_z)^T$ and $g^l = (0\ 0\ 1)^T$. Therefore, we have

$$\theta = \operatorname{atan2}\left(g_x,\ \sqrt{g_y^2 + g_z^2}\right) \tag{14}$$

This equation is actually the same as equation (10) if the sensor module is not moving. Using equation (13) and (14), we obtain

$$\phi = \text{atan}\,2\left(g_y\,\text{sign}(\cos\theta), g_z\,\text{sign}(\cos\theta)\right) \tag{15}$$

This equation is different from equation (11). Hence the ϕ angle of roll is not the y-tilt angle β. y-tilt should not be used as roll for orientation calculation and magnetometer compensation.

After both pitch θ and roll ϕ are found from acceleration data, the attitude, or yaw ψ, can be found from magnetic field data by solving the following simultaneous equation:

$$\mathbf{m}^b = \mathbf{R}^T \cdot \mathbf{m}^l \tag{16}$$

where $\mathbf{m}^b = [m_x\ m_y\ m_z]^T$, $\mathbf{m}^l = [a\ 0\ b]^T$ (a and b are the horizontal and vertical components of the magnetic field of the earth vector) and $\mathbf{R}^T = \mathbf{R}_1^T(\psi,z)\mathbf{R}_2^T(\theta,y)\mathbf{R}_3^T(\phi,x)$. Hence we get

$$\mathbf{R}_1^T(\psi,z) \cdot \mathbf{m}^l = \mathbf{R}_2^T(\theta,y) \cdot \mathbf{R}_3^T(\phi,x) \cdot \mathbf{m}^b \tag{17}$$

As illustrated in Fig. 6, h_x, h_y and h_z are the projections of magnetic field on the horizontal plane and the z-axis (pointing downward) respectively, i.e., $\mathbf{h}^b = [h_x\ h_y\ h_z]^T = (a\cos\psi, -a\sin\psi]^T$. Then the compensated magnetic field can be obtained as below:

$$h_x = m_x \cos\theta + m_y \sin\theta\sin\phi + m_z \sin\theta\cos\phi \tag{18}$$

$$h_y = m_y \cos\phi - m_z \sin\phi \tag{19}$$

$$h_z = m_x \sin\theta + m_y \cos\theta\sin\phi + m_z \cos\theta\cos\phi \tag{20}$$

Accordingly, the heading on the horizontal plane, or the Euler angle yaw, can be defined as

$$\psi = \text{atan}\,2\left[-h_y, h_x\right] \tag{21}$$

The angle ψ calculated in equation (21) is actually a compass reading, i.e., the heading relative to magnetic north. To get the true heading, denoted as ψ_T here, the magnetic declination D must be deduced from the compass reading ψ:

$$\psi_T = \psi - D \tag{22}$$

where D is the difference between magnetic north and true north. This difference is caused by the tilt of the earth's magnetic field generator relative to the earth's spin axis.

5. Experimental study

To implement the methods introduced in the Section 4,. A commercial sensor module shall be carefully selected And then mounted to on a person rigidly. The result of the on-site pedestrian tracking is promising.

5.1 Sensor module selection

Many factors, including size, cost, dynamic range, sampling rate and bias, have to be considered in choosing an appropriate sensor module. Firstly, the size of the sensor module should be small enough to be installed unobtrusively on footwear. Consequently, all 3D gyroscopes, 3D accelerometers and 3D magnetometers should be integrated into, actually orthogonally mounted in, this inertial/magnetic sensor module. Secondly, the price of this sensor module should be low enough to meet budget constraints for the targeted applications. Thirdly, the dynamic range of the sensor module should meet the requirements of the applications. Fourthly, the sampling rate should be high enough to avoid sampling errors. The sampling rate of the sensors should be at least 100 Hz. Finally, because of the sensitivity of sensors to temperature, each sensor in a module should be individually compensated for bias, which is a function of temperature.

With all these considerations an inertial measurement unit (IMU) was selected to build the prototype for testing. The IMU provides serial digital outputs of 3D acceleration, 3D angular velocity, and 3D magnetic field.

5.2 Sensor module mounting

In the experiments, the sensor module is mounted on the arch, i.e., the upper of closed footwear, as shown in Fig 7. The installation of the sensor should ensure that the relative movement between the sensor module and the foot-ware is minimized.

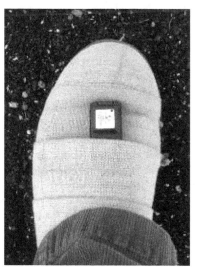

Fig. 7. Sensor Module Mounting

5.3 Field experiment

Using the methods introduced in Section 4 and the sensor module selected in Section 5.1, several experiments have been conducted at different paces in different places and different times. Most of these experiments are two-dimensional (2D) tests.

Fig. 8. Tracking the walk on a J-shaped path at a park

In the setup of experiments related to Fig 8, the walk is along an outdoor J-shaped path near a playground. The plotted trace recorded is similar to the actual path walked.

6. Sensor fusion and Kalman filtering

6.1 Sensor fusion

The use of higher quality inertial/magnetic sensor modules will improve positioning accuracy but the cost of such a solution is often much higher. An alternative solution is to use various positioning data from multiple sources such as radio navigation aids and additional sensors onboard. Through the control applied to the drift error, a positioning system using multiple data sources "fused" together is able to outperform a system using a single data source. Complementary filtering techniques were used to integrate data from different positioning aids in the past. Nowadays, Kalman filtering is popular in combination of two estimates of a variable. For example, IPS data can be integrated with GPS data to provide enforced positioning ability in both indoor and outdoor environment.

6.2 Kalman filter

Although dynamical behavior of a linear system is often expressed in the form of continuous differential equations, in practice, the measurements are provided at discrete intervals of time. The form of difference equations are used as below:

$$\mathbf{x}_k = \mathbf{F}_{k-1}\mathbf{x}_{k-1} + \mathbf{B}_{k-1}\mathbf{u}_{k-1} + \mathbf{w}_{k-1} \tag{23}$$

where the n vector, $\mathbf{x}_k$, is called the state of the system at time k, $\mathbf{u}_{k-1}$ is a p vector of deterministic inputs and $\mathbf{w}_{k-1}$ is the system noise at time k-1. $\mathbf{F}_{k-1}$ is an nxn state transition matrix from time k-1 to time k, known as the system matrix, and $\mathbf{B}_{k-1}$ is an nxp system input matrix. $\mathbf{F}_{k-1}$ and $\mathbf{B}_{k-1}$ are constant or time varying matrices. The system (or process) noise $\mathbf{w}_k$ is normally distributed with zero mean and a power spectral density of $\mathbf{Q}_{k-1}$.

The vector $\mathbf{x}_k$ contains all the information regarding the present state of the system, but cannot be measured directly. Instead, only the system's m vector, $\mathbf{z}_k$, is available. This measurement vector is a linear combination of the state, $\mathbf{x}_k$, that is corrupted by the measurement noise $\mathbf{v}_k$. This can be expressed in terms of the system state by the following equation:

$$\mathbf{z}_k = \mathbf{H}_k\mathbf{x}_k + \mathbf{v}_k \tag{24}$$

where $\mathbf{H}_k$ is an mxn measurement matrix. The measurement (or observation) noise, $\mathbf{v}_k$, has zero-mean and is normally distributed, with power spectral density $\mathbf{R}_k$.

The Kalman filter for the system described here seeks to provide the best estimates of the states, $\mathbf{x}_k$, using the measurements, $\mathbf{z}_k$, model of the system provided by the matrices $\mathbf{F}_k$, $\mathbf{B}_k$ and $\mathbf{H}_k$, and knowledge of the system and measurement statistics given in the matrices $\mathbf{Q}_k$ and $\mathbf{R}_k$.

From (23) and (24), we can recognize that Kalman filter is a recursive filter: only the state of last time step and the current measurement are used. The estimation is conceptualized as two distinct phases, "Predict" and "Update", described by two distinct set of equations. In the first phase, the prediction is based on previous best estimate. The result is called a priori state estimate. In the second phase, the updating is done based on the a priori state estimate with new measurement. The result is a posteriori state estimate. The two set of distinct equations are as below:

The prediction process

Predicted (a priori) state estimate is the best estimate of the state at time k-1, denoted as $\mathbf{x}_{k-1/k-1}$. Since the process noise, $\mathbf{w}_{k-1}$, is normally distributed with zero-mean, the best prediction of is given by

$$\mathbf{x}_{k/k-1} = \mathbf{F}_{k-1}\mathbf{x}_{k-1/k-1} \tag{25}$$

while the predicted (a priori) estimate covariance at time k predicted at time k-1, is given by

$$\mathbf{P}_{k/k-1} = \mathbf{F}_{k-1}\mathbf{P}_{k-1/k-1}\mathbf{F}_{k-1}^{\mathrm{T}} + \mathbf{Q}_{k-1} \tag{26}$$

The measurement update

On the arrival of new measurement $\mathbf{z}_k$, at time k, it is compared with the a priori state estimate of the measurement. The measurement is then used to update the prediction to generate a best estimate, i.e., the a posteriori state estimate at time k-1:

$$\mathbf{x}_{k/k} = \mathbf{x}_{k/k-1} - \mathbf{K}_k[\mathbf{H}_k\mathbf{x}_{k/k-1} - \mathbf{z}_k] \tag{27}$$

Accordingly, the updated (*a posteriori*) estimate covariance is:

$$\mathbf{P}_{k/k} = \mathbf{P}_{k/k-1} - \mathbf{K}_k \mathbf{H}_k \mathbf{P}_{k/k-1} \tag{28}$$

where the optimal Kalman gain matrix is given by:

$$\mathbf{K}_k = \mathbf{P}_{k/k-1} \mathbf{H}_k^\mathrm{T} \left[\mathbf{H}_k \mathbf{P}_{k/k-1} \mathbf{H}_k^\mathrm{T} + \mathbf{R}_k \right] \tag{29}$$

Typically, the two phases alternate, with the prediction advancing the state until the next scheduled measurement, and the update incorporating the measurement.

6.3 Extended Kalman filter

Kalman filter is optimal in either the least square sense or maximum likelihood sense if it is applied to linear dynamic system with zero mean, normally distributed noise. Kalman filter becomes not optimal when the system is nonlinear, or the noise is not normally distributed.

EKF is the nonlinear version of the Kalman filter. EKF linearizes the dynamic system about the current mean and covariance. The estimate is suboptimal, i.e. it is an approximation of the optimal estimate.

In EKF, the state transition and measurement models may be non-linear differentiable functions:

$$\mathbf{x}_k = f\left(\mathbf{x}_{k-1}, \mathbf{u}_{k-1}\right) + \mathbf{w}_{k-1} \tag{30}$$

$$\mathbf{z}_k = h(\mathbf{x}_k) + \mathbf{v}_k \tag{31}$$

where system noise $\mathbf{w}_k$ is normally distributed with zero mean and covariance $\mathbf{Q}_k$ while measurement noise $\mathbf{v}_k$ is zero mean normally distributed with covariance $\mathbf{R}_k$. The functions f and h can be used to estimate the current state and measurement from previous state and measurement respectively. However, the current covariance of system noise and measurement noise cannot be estimated from these two functions directly. Instead the Jacobians, matrices of partial derivatives, must be calculated. The state transition and measurement matrices are defined as:

$$\mathbf{F}_{k-1} = \left.\frac{\partial f}{\partial \mathbf{x}}\right|_{\mathbf{x}_{k-1/k-1}, \mathbf{u}_{k-1}} \tag{32}$$

$$\mathbf{H}_k = \left.\frac{\partial h}{\partial \mathbf{x}}\right|_{\mathbf{x}_{k/k-1}} \tag{33}$$

Similar with Kalman filter, each EKF iteration is also composed of two distinct phases, representing by two set of equations as below:

The prediction process

Predicted state estimate and its covariance through linearization of system dynamics and application of the prediction step of the Kalman filter

$$\mathbf{x}_{k/k-1} = f\left(\mathbf{x}_{k-1/k-1}, \mathbf{u}_{k-1}\right) \tag{34}$$

$$\mathbf{P}_{k/k-1} = \mathbf{F}_{k-1}\mathbf{P}_{k-1/k-1}\mathbf{F}_{k-1}^{T} + \mathbf{Q}_{k-1} \tag{35}$$

The measurement update

Measurement residual is calculated through linearization of system dynamics

$$\mathbf{y}_k = \mathbf{z}_k - h\left(\mathbf{x}_{k/k-1}\right) \tag{36}$$

The relevant covariance is

$$\mathbf{S}_k = \mathbf{H}_k\mathbf{P}_{k/k-1}\mathbf{H}_k^{\mathrm{T}} + \mathbf{R}_k \tag{37}$$

The suboptimal Kalman gain is

$$\mathbf{K}_k = \mathbf{P}_{k/k-1}\mathbf{H}_k^{\mathrm{T}}\mathbf{S}_k^{-1} \tag{38}$$

The state estimate is updated as

$$\mathbf{x}_{k/k} = \mathbf{x}_{k/k-1} + \mathbf{K}_k\mathbf{y}_k \tag{39}$$

The relevant estimate covariance is also updated as

$$\mathbf{P}_{k/k} = \left(\mathbf{I} - \mathbf{K}_k\mathbf{H}_k\right)\mathbf{P}_{k/k-1} \tag{40}$$

6.4 Application examples

Kalman filter, sometimes with other techniques, have already been used for data fusion and positioning. Recently, Zhao & Wang (2011) use EKF for the fusion of data from inertial sensors, ultrasonic sensors, and magnetic sensor. A 3D magnetic sensor and a 3D accelerometer are combined to measure the gravity and the earth's magnetic field for the static orientation. A 3D gyroscope is used to obtain the dynamic orientation. The integral drift of accelerometer is periodically calibrated by an ultrasonic sensor. All these data are fused by EKF. The measurements are the position obtained by ultrasonic sensor and the orientation obtained by magnetic sensors and accelerometers. The states incorporated the position, velocity, and orientation. The experimental results demonstrated that the uncertainty of orientation and position is lowered.

In another research, Shen et al (2011) apply Kalman filter, together with Fast Orthogonal Search (FOS) in the 2D navigation for land vehicles. Kalman filter is used for data fusion of GPS and inertial sensors, i.e. the integration of GPS, IMU and vehicle built-in odometer. FOS is a nonlinear error identification technique. It is used to model non-stationary stochastic sensor errors and non-linear inertial errors in the study. The road test trajectories confirm that a module with both Kalman filtering and FOS technique outperforms that with Kalman filtering only.

It can be concluded that improvement of positioning accuracy will be possible if data from multiple sensors, including magnetic sensors, are fused by Kalman filter. Further improvement is achievable if Kalman filtering is combined with other techniques.

7. Summary

In this chapter, we firstly defined local positioning systems and introduced their applications. Then we presented strap-down inertial positioning techniques covering not only the concepts of dead reckoning and inertial position, but also the algorithms, especially for the rotation of reference frames and the tracking of the orientation of a moving object. We also evaluated the performances of inertial sensors, i.e., gyroscopes and accelerometers and pointed out that the increasing integral drift over time is the main problem affecting the accuracy of positioning. As an example among the efforts to solve this problem, we proposed an approach to integrate pedestrian dead reckoning with inertial positioning in precise tracking, where the characteristics of human gaits are utilized. Magnetic sensors have to be incorporated to update the orientation of moving objects and the relevant algorithm is provided. The experimental study was also included to demonstrate the implementation and the result of positioning. Finally, among existing solutions to enhance the accuracy of local positioning, Kalman filter, EKF and their algorithms are introduced and the examples of their applications are also presented. Magnetic sensors are important resources of data to be fused through Kalman filtering.

8. References

Huang, C.; Liao, Z. & Zhao, L. (2010). Synergism of SIN and PDR in Self-contained Pedestrian Tracking. *IEEE Sensor Journal*, Vol. 10, No. 8, pp1349-1359.

McBrewster, J.; Miller, F. P. & Vandome,A. F. (2009). *Navigation: Latitude, Longitude, Dead Reckoning, Pilotage, Celestial Navigation, Marine Chronometer, Sextant, Inertial Navigation System, Radio Navigation, Satellite System, Passage Planning*. Alphascript Publishing.

Mezentsev, O.; Lachapelle, G. & Collin, J. (2005). Pedestrian Dead Reckoning – A Solution to Navigation in GPS Signal Degraded Areas. *Geomatica*, Vol. 59, No.2, pp. 175-182.

Rose, J. & Gamble, J. (1994). *Human Walking*, 2nd Ed. Williams & Wilkins.

Sarton, G. & George, A. (1959). *History of Science: Hellenistic Science and Culture in the Last Three Centuries B.C.* New York: The Norton Library, Norton & Company Inc..

Shen, Z.; Georgy, J.; Korenberg, M. J. & Noureldin, A. (2011). Low Cost Two Dimension Navigation Using an Augmented Kalman Filter/Fast Orthogonal Search Module for the Integration of Reduced Inertial Sensor System and Global Positioning System. *Transportation Research. Part C, Emerging Technologies*, vol. 19, no. 6, pp.1111-1132.

Stirling, R. & Fyfe, K. (2005). Evaluation of a New Method of Heading Estimation for Pedestrian Dead Reckoning Using Shoe Mounted Sensors. *The Journal of Navigation*, vol.58, pp. 31-45, The Royal Institute of Navigation.

Thiessen, F. J. & Dales, D. N. (1983). *Automotive Steering, Suspension and Braking Systems*, Reston Pub. Co.

Woodman, O. & Harle, R. (2008). Pedestrian Localisation for Indoor Environments. *Proceedings of the Tenth International Conference on Ubiquitous Computing (UbiComp 08), Seoul, Korea*, ACM, Sep 2008.

Zhao, H. & Wang, Z. (2011). Motion Measurement Using Inertial Sensors, Ultrasonic Sensors, and Magnetometers with Extended Kalman Filter for Data Fusion, *IEEE Sensors Journal*, vol. pp, no. 99, pp. 1-8.

6

Magnetic Sensors for Biomedical Applications

Guillermo Rivero, Marta Multigner and Jorge Spottorno
Instituto de Magnetismo Aplicado (Universidad Complutense de Madrid)
Spain

1. Introduction

The aim of this chapter is to give an overview of the applications of magnetic and magnetoelastic sensors and actuators in Biomedicine.

Over the last century, life expectancy on our planet has experienced a remarkable increase. That is if we exclude those regions ravaged by war and those others in which the population lives, or more probably survives, under conditions that we would not tolerate. In Europe this increase has been spectacular. Average life expectancy has reached 80 years old among the male population and even more for the female population and the tendency is upward. However, this has caused important health problems, as our bodies are not usually prepared to function for such a long time without repairs and/or replacements.

As a result, so-called substitution surgery is now taking up an increasing percentage of the operations performed in hospitals; such as bone prostheses for hips, knees and teeth, implants for eyes and ears and artificial sphincters and penises. Besides this, it is necessary to make periodic analyses in order to test that the levels of glucose, and cholesterol, etc. are correct. So, the application of magnetism in Biomedicine covers a wide field of devices, from sensors to determine the concentration of several elements in the blood to prostheses, both active and passive, for replacing organs or articulations.

Apart from all the above, some experimental treatments also use magnetic sensors or actuators. For example, magnetic nanoparticles are used for hyperthermia treatments against tumour cells or for drug delivery. In addition, much more sophisticated and expensive treatments, like magneto encephalography techniques, use magnetic elements.

In this chapter we are going to describe a small number of devices, some of which are now being researched. They are divided in two categories: magnetic sensors and magnetic actuators, according to the following scheme:

- Magnetic Sensors
  - In situ measurement of the mass evolution of cell culture
  - Test of blood coagulation
  - Sensor system for early detection of heart valve bio prostheses failure
- Magnetic Actuators
  - Magnetic endoluminal artificial urinary sphincter

- Hyperthermia HeLa cell treatment with silica-coated manganese oxide nanoparticles

In the development of these magnetic devices, as well as others not described here, there are some aspects that are common to the magnetic devices used in other very different fields as, for example, railway sensors, automotive sensors or aerospace vehicles. These may apply the basic magnetic properties or the signal acquisition and control process. Nevertheless, there are some aspects, as for instance, biocompatibility, corrosion resistance, size limitation and patient comfort, in the case of a human implant, that are specific to this application. We will try to describe them through our work in this area.

2. Magnetic sensors

Magnetic sensors are based, with very few exceptions, on the same working principle: the change of magnetic moment that takes place in a magnetic material, (normally ferromagnetic), when submitted to a magnetic field, a change of temperature or, in the case of magnetoelastic materials, to a mechanical stress, generated by the medium, in which they are immersed. The measurement of this variation makes it possible to study the changes that occur in the surroundings of the sensor.

In the case of the biomedical applications, the medium may be a fluid like, for instance, blood, the cerebrospinal fluid or a culture medium or it may be organic tissue. In any case there will be an interaction between both elements, the magnetic material and the organic medium, which is undesirable. On the one hand, the organic media may be harmed since the most widely used magnetic materials, (transition metals and their alloys), are not biocompatible, except for iron which is very biocompatible; (human organs and tissues have a great affinity for iron). On the other hand, the above mentioned organic fluids have Cl^-, $Ca^=$, Na^+, K^+ ions and several free organic radicals which are, in general, corrosive for magnetic materials. The solution to this problem consists in coating the magnetic material with a thin layer (100 to 200 nm) of another biocompatible material, like gold, platinum, titanium oxide, silica or alumina. The most widely used coating methods are evaporation, sputtering, electrolysis and ionic implantation. The non metallic parts of the sensor also have to be made of biocompatible polymers or ceramics such as Teflon, medical silicones, and silica.

Another aspect that has to be taken into account regarding these sensors is the transmission and monitoring of the signal. Wireless transmission, which is one option in other application fields, is a compulsory one in biosensors. This is for obvious reasons in the case of internal implants, while in the case of the external ones wireless transmission is convenient so as not to restrict the patient's movement. Magnetic sensors and actuators offer a great advantage from the start: the magnetic field is itself a wireless transmission. The advances that have taken place in the past decades in the field of cellular phones, combined with magnetic sensors and biosensors, open the door to the final development of "telediagnosis".

The sensors that are presented in this section are based on the magnetoelastic properties of several magnetostrictive ferromagnetic alloys (Chikazumi, 1964). As is well known, a magnetostrictive material is characterized by a change of its macroscopic dimensions when its magnetization changes, getting longer or shorter (positive or negative magnetostriction respectively) along the direction of the magnetization. In other words, the energy of the magnetizing field is used not only in the magnetization process but also in elastically deforming the material. The opposite phenomenon also takes place. When the material is deformed due to an external stress, not all the mechanical work is used in deforming the material, a part of the stress is used in magnetizing the material. In the free energy of the solid, a third term has to be added to the terms of elastic and magnetic energy that is called the magnetoelastic energy term. It takes into account the energy transfer between the elastic system and the magnetic one:

$$F = F_{el}(\varepsilon) + F_{mg}(m) + F_{mgel}(m, \varepsilon) \tag{1}$$

As a consequence, in these materials the magnetic susceptibility changes with the deformation (and consequently with stress) of the material and, inversely, the elastic constant of the material changes with its magnetization (and thus the applied magnetic field). The magnetic coupling term, being the one that takes into account the energy exchange between both systems, shows a resonance phenomenon: it has a maximum for a particular frequency that depends on the magnetic and elastic properties of the material.

The materials that are most widely used are the amorphous magnetostrictive alloys, in which the absence of magnetocristalline anisotropy makes it easy to control the resonance frequency, acting on the magnetic anisotropy of the material by means of appropriate thermal treatments. The amorphous magnetic materials with positive magnetostriction constant, made by the Taylor technique (Marín & Hernando, 2004), are well known due to their availability for use in wireless sensors based on magnetoelastic resonance (Vázquez & Hernando, 1996). Some recent works show the possibility of using these materials as magnetoelastic biosensors(Shem et al., 2009; Xie et al., 2009). Other works show a detailed study of the magnetoelastic coupling phenomenon in ribbons (Hernando 1983). Recently, there has been published a development, for monitoring in situ the mass of a cell culture, based on the magnetoelastic resonance of ribbon shaped metallic glasses (Rivero et al., 2008). On the other hand, magnetic amorphous microwires coated with Pyrex have attracted much interest due to their particular properties and their simple manufacturing process based on the Taylor technique. The use of these materials can reduce the dimensions of the set-up and the Pyrex cover makes them appropriate for many applications in the medical field.

2.1 *In situ* measurement of the mass evolution of cell culture

Cell cultures constitute one of the most frequently used assays in biology and also one of the diagnostic methods most used in Medicine. The growing of several microorganism strains, like cells, bacteria and virus etc., under conditions in which atmosphere, temperature, humidity and nutrients are controlled have been studied.

On the other hand, the research on the treatment of tumours by hyperthermia, in the last decade, has been focused on the different behaviour of normal and tumour cells over temperature. Generally, normal cells show better temperature resistance than the tumour ones. Consequently, it is very important to determine the optimum temperature for hyperthermia treatments.

The monitoring of the progress of a cell culture is conventionally done by direct observation of the evolution that takes place on a culture plate, (the support on which the cells are placed), using a microscope. It is essential to have perfect control over the microenvironment in which the culture is developing to ensure that any change in the behaviour and multiplication of the cells is only ascribable to the cell behaviour in a given situation and not to poor culture conditions. For this reason, the cultures are carried out inside incubation chambers where it is possible to control factors such as humidity, pH and temperature. Among all these factors, probably the most important one is strict control of the temperature since it is a critical factor in the majority of the experiments performed with cells, particularly in mammalian ones. However when the evolution of the cells is observed using a microscope, it is necessary to extract the cells from the chamber, submitting them with what may result in undetermined damages due to changes in temperature, aside from the mechanical damages that take place in each measurement.

So a system that makes it possible to continuously quantify the mass evolution of a cellular culture "in situ", that is, without extraction from the incubation chamber would be very useful and to our knowledge, it does not exist. This sensor would enable the cellular growth to be studied accurately under different temperatures, without the periodical mechanical damage and the cooling-heating processes produced by the extraction and movement of the culture from the chamber to the microscope where the counting of the cells is performed.

One solution is a system sensor based on the magneto-elastic resonance of ferromagnetic amorphous ribbons. Other systems based on the same principle have been developed before for detecting other biological agents (Lakshmanan et al., 2007; Wan et al., 2007)

The system consists of the following elements (Fig.1):

- A culture plate that has been designed with two separate baths, each one containing an amorphous magnetoelastic ribbon with an iron based composition of area 40 x 4 mm^2 and 15 microns in thickness, coated with 250 nm of TiO_2 immersed in a culture medium. Only one bath is seeded with cells. So we have a continuous reference independent of temperature.
- An arrangement of two coils and a permanent magnet under the plate to apply the bias and the alternating field.
- A scanner impedance meter (Fig. 2), connected to the coils, which measures the coils impedance around the resonance frequency of the ribbons.
- A software program developed to extract the ribbons resonance frequency from the impedance measurements.

The working of the device is as follows. Inside the culture plate designed for the device two ribbons of an amorphous magnetostrictive material with the same dimensions are placed in the compartments designed for them immersed in the culture environment. One of the ribbons acts as a reference, in order to evaluate the changes on the magnetoelastic resonance frequency due to possible changes of temperature or other factors that are not directly related to the evolution of the cell culture. So, it is immersed in one of the baths containing the culture environment but without cells. A seed of the cells, whose evolution and growth is being studied, is placed in the bath containing the other ribbon. By means of the coils system (excitation-pick up system) an electromagnetic field of frequency ω is applied over the whole set-up and the variations in the magnetic flux density, created by the sensor system, are collected. The designed electronic system makes it possible to change the frequency of the applied electromagnetic field as well as receiving the answer of the sensor in function of the frequency.

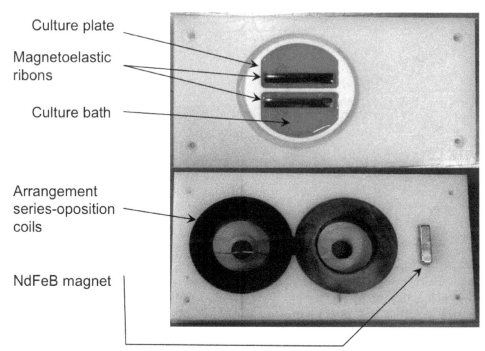

Fig. 1. Experimental set-up to measure the mass evolution of cell culture.

To perform the measurement an electronic system is used. Its flowchart is shown in figure 2. By means of a DDS (1) (Direct Digital Synthesizer) device the values of two sinusoidal signals in quadrature, $sin(\omega t)$ and $cos(\omega t)$ are obtained. By means of (2) a fixed voltage difference is generated that is used by the D/A (Analogical/Digital) (3) and A/D (8) converters. The sinusoidal signal $sin(\omega t)$ generated by (1) is transformed into an analogical signal by the D/A converter. This signal is amplified by means of a power amplifier (4). The amplified signal feeds the excitation/pick up system (5) that is made up by two opposite series connected coils, which generate an alternating magnetic field that energizes the magnetostrictive ribbons (6) and that also enables answer from the magnetoelastic sensor to be picked up.

The variations in the induced voltage created in the pick up system are converted into electric current variations by the voltage/current converter (7). By means of the analogical/digital converter we can obtained the digital measurement of the electric current that flows through the exciting coils. The digital signal processor (9) permits us to extract the in phase and in quadrature components of the current measured by (8) and by (10) we obtain, for each frequency, the magnitude of the impedance of the coil system, which makes it possible to measure the resonance frequency of the coils / magnetoelastic sensor setup.

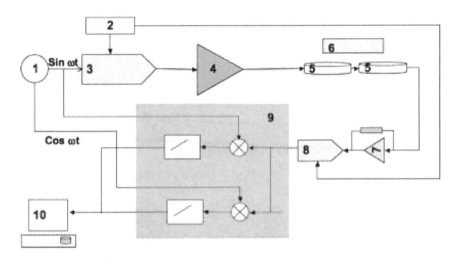

1) Direct Digital Synthesizer; 2) Voltage reference; 3) D/A Converter; 4) Power amplifier; 5) Exciting/Pick up coils; 6) Magnetoelastic sensor; 7) Voltage/Current Converter; 8) A/D converter; 9) Digital signal processor; 10) Acquisition

Fig. 2. Flowchart of the scanning impedance meter:

When the frequency of the electromagnetic field matches the frequency of the magnetoelastic resonance of any of the ribbons, the transformation of magnetic energy into an elastic one reaches a maximum. This resonance is detected by the coil system, showing a characteristic peak in the measurement of the coil impedance. The change of the magnetic resonance frequency when the mass of a magnetostrictive material changes by Δm is given by (Grimes et al, 1999):

$$\Delta\omega = -\omega\frac{\Delta m}{2m} \tag{2}$$

Where ω is the initial resonance frequency, m is is the initial mass, Δm is the change of mass and $\Delta\omega$ is the resonance frequency variation. Thus, the resonance frequency of a magnetoelastic ribbon decreases when the mass on the ribbon increases. Before the cells are seeded, both ribbons have the same resonance frequency and only one resonance peak is detected. When the culture cells on the seeded ribbon grow, two different peaks appear, corresponding to both ribbons, as shown in figure 3. The frequency interval between the peaks determines the amount of culture mass.

Preliminary experiments have been made with a human cervical cancer cell line (HeLa) in Dulbecco's modified Eagle medium, supplemented with 10% fetal bovine serum (FBS) and 1% (v/v) antibiotic solution. Changes in the mass culture of less than 1% have been detected.

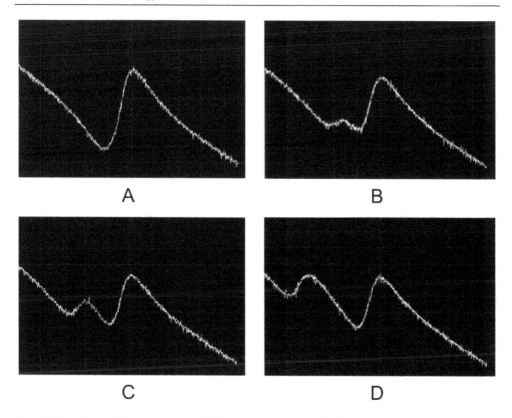

Fig. 3. Evolution of the resonance of ribbons arrangement. In the initial state A, both ribbons have the same resonance frequency. The left peak increases and decreases its frequency with the amount of mass on the corresponding ribbon.

2.2 Test of blood coagulation

The Time of Prothrombine (TP), and the values that derive from it, like the International Normalized Ratio (INR), are used for determining the tendency of the blood to coagulate in the presence of possible biological disorders like hepatic failure or K vitamin deficiency. It is also used to control patients that take anticoagulation drugs like warfarina or acenocumarol to prevent coronary thrombosis processes in cardiac pathologies: atrial fibrillation, and auricular fibrillation, etc. In addition, due to the increase in life expectancy, an ever growing percentage of the population needs to replace one or more cardiac valves with a mechanical prosthesis and, in some cases, more than once. All these people must be submitted to a life-long treatment using anticoagulants. This entails a permanent risk for them. So it is not rare that the problem of controlling the appropriate level of the anticoagulation agent in the treatment of several pathologies may affect about 1 – 2 % of people in the European Community. The Food and Drugs Administration (FDA) of USA estimates that about two million people begin a treatment with warfarin every year. These patients need to be controlled periodically due to the side effects of the treatment. These controls must be done in hospital. Nowadays, there are some portable devices (Neel et al., 1998; Askew et al., 2010) that enable the control to be done more conveniently, but the results are not very reliable. The main causes of error are usually the influence of temperature and the amount of blood involved in the test. Other methods based on the measurement of blood viscosity (Drobrovol'skii et al., 1999) need a large amount of blood.

A sensor, based on a magnetoelastic material, can be used to determine the TP and the INR in patients under anticoagulation treatment, without the need for specialized staff or installations. The method is based on the variation in the magnetic permeability of a magnetoelastic microwire induced by the change on the blood viscosity when it coagulates. When the blood coagulates, the viscosity force applied to the immersed wire dissipates, as heat, a portion of the magnetic energy supplied by the magnetic field. Therefore the apparent magnetic permeability of the microwire decreases due to the magnetoelastic coupling. The experimental set-up compares this permeability with that of a reference wire immersed in an inalterable fluid. The absolute value of the measured signal tends to a maximum when the blood coagulates. The time to raise this maximum enables the TP and the INR to be calculated.

The sensor consists of two identical microwires, with an iron base, that are placed into two capillaries with tenths of mm. inner diameter and around 5 cm in length. These capillaries are surrounded by one coil each, which are fed by two power amplifiers driven by a signal generator. The capillaries are filled with blood and with a fluid of reference, respectively. The difference between both signals increases when the blood coagulation process begins and its absolute value tends to a maximum when the blood is fully clotted. The scheme of the experimental set-up used to measure the blood coagulation is illustrated in figure 4.

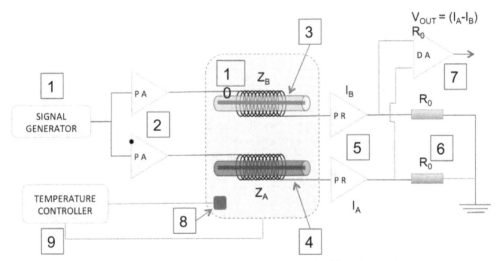

Fig. 4. Scheme of the experimental setup used to measure the blood coagulation.

The sensor system consists of the following components (see figure):

- A signal generator that provides a sinusoidal voltage signal in a range of 10 – 50 KHz. (1)
- Two power amplifiers, driven by the signal of the generator, which feed two equal coils with the microwire arrangements. The gain of amplifiers can be adjusted to obtain zero signal before the test. (2)
- Coil with reference arrangement (capillary + microwire + fluid of reference) (3)
- Coil with measurement arrangement (capillary + microwire + blood) (4)
- Two precision rectifiers (5)
- Two non-inductive resistors (6)
- A differential amplifier that gives the difference between the rectified voltages in the non-inductive resistors. (7)
- A temperature controlled chamber in which the sensor system is introduced to perform the test at constant temperature of 25 °C.
- A temperature sensor (8) that measures the temperature in the temperature controlled chamber. The signal of the temperature sensor is sent to a temperature controller device (9) that keeps the chamber temperature at 25 °C.

Tests with blood samples that coagulate at different rates are illustrated in figure 5.

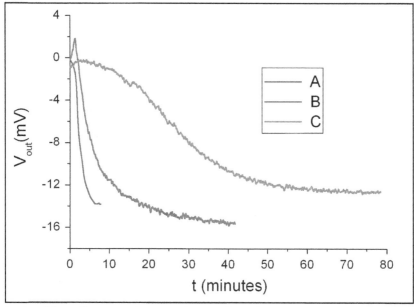

Fig. 5. Study of the influence in the coagulation of blood with different amounts of coagulant agent. The amount of blood is 200 µl, and was mixed with 5µl of coagulant (curve A), 1µl of coagulant (curve B), and without coagulant (curve C).

It can be observed that blood, in the presence of the greatest quantity of coagulant agent, coagulates in a time 10 times faster than the blood without the coagulant agent. The output potential, V_{OUT}, of the differential amplifier shown in figure 4 is the difference between V_{REF} (Voltage difference in the circuit of the reference coils) and V_{BLOOD} (Voltage difference in the circuit we are measuring the blood in). Both are RL circuits and the inductance L and resistance R_L:

$$L = \mu_0 \frac{N^2}{l}(S_C + \mu_W S_W) \qquad\qquad R_L = \frac{2N\sqrt{\pi S_C}}{\sigma_{Cu} S_{Cu}} \qquad\qquad (3)$$

Being μ_W the relative permeability of the microwire, μ_0 the permeability of free space, N the number of turns, l the length of the coil, σ_{Cu} the copper conductivity and S_C and S_W the sections of the coil and the microwire, respectively.

Since in this setup, when R_0 and $R_L << w L$ then:

$$V_S = V_0 \, e^{i\omega t} \quad\Rightarrow\quad V = V_0 \frac{R_0}{\omega L} e^{i(\omega t - \pi/2)} \qquad\qquad (4)$$

Once they are rectified, the voltage differences are:

$$V_{REF} = V_0 \frac{R_0\sqrt{2}}{\omega L_{REF}} \qquad and \qquad V_{BLOOD} = V_0 \frac{R_0\sqrt{2}}{\omega L_{BLOOD}} \qquad\qquad (5)$$

Where L_{REF} and L_{BLOOD} are the inductances of the reference and measurement coils, respectively. Thus the output of the differential amplifier (see figure 1):

$$V_{OUT} = V_{REF} - V_{BLOOD} = V_0 \frac{R_0\sqrt{2}}{\omega}\left(\frac{1}{L_{REF}} - \frac{1}{L_{BLOOD}}\right) \qquad\qquad (6)$$

This output is taken to zero at the initial moment, adjusting the gain of the feeding amplifiers. From this moment, the permeability, μ_W, of the microwire immersed in blood begins to decrease because of the effect of the viscosity force the blood exerts on it while it clots. This makes L_{BLOOD} decrease as well.

The output of the differential amplifier depends on time as follows:

$$V_{OUT}(t) = V_0 \frac{R_0\sqrt{2}}{\omega}\left(\frac{1}{L_{BLOOD}(0)} - \frac{1}{L_{BLOOD}(t)}\right) = \frac{V_0 R_0 \sqrt{2}}{\omega L_{BLOOD}(0)}\left(1 - \frac{S_C + S_W \mu_W(0)}{S_C + S_W \mu_W(t)}\right) \qquad (7)$$

Figure 6 shows the curve obtained from human blood without coagulant agent and the theoretical $V_{OUT}(t)$ curve, following the equation (7). In this expression the best adjustment to experimental results is obtained assuming a relative permeability change of the microwire during the coagulation process as:

$$\mu_W(t) = \mu_{W,\infty} + \left(\mu_{W,0} - \mu_{W,\infty}\right)\exp\left(-\frac{t^2}{\tau^2}\right) \qquad\qquad (8)$$

Being $\mu_{W,0}$ and $\mu_{W,\infty}$ the measured values of the initial permeability and the permeability when the blood is completely coagulated, respectively, and τ a time constant that depends on the coagulation properties of the blood sample. The value of τ for the best adjustment to the experimental values enables the TP and INR to be determined.

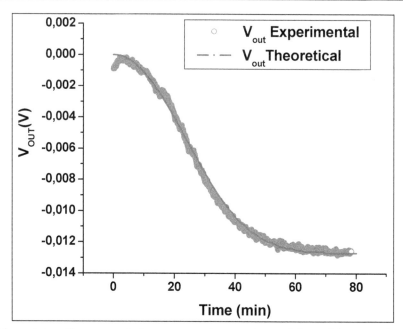

Fig. 6. The points with a green circle represent the experimental data of the coagulation of blood without coagulant. The line in red represents the proposed approximation adjusted to the experiment.

2.3 Sensor system for early detection of heart valve bioprostheses failure

Heart valves are widely used in cardiac diseases. For example in 2002 in Spain 9269 heart valves were implanted as against 310 heart transplants. There are two kinds of heart valves, mechanical and biological ones. Biological heart valves, currently made with calf pericardium, have a similar shape to that of the original ones and better hemodynamic conditions than the mechanical ones. Moreover they have the advantage that the patient does not need lifelong treatment with anticoagulants, as happens with the mechanical valves. On the other hand, biological cardiac valves, or bioprostheses, have the inconvenience of their limited durability (about 10 years on average). What is more troublesome is the fact that they have an important dispersion that may be estimated in plus/minus 3 years (Kouchoukos et al, 2003).

These bioprostheses fail for several reasons, similar to those of the original ones and there are many factors that may contribute to their deterioration. The most important are biochemical degradation and mechanical damage of the tissue. Mechanical fatigue is the result of the great number of opening and closing cycles (approximately 30 million per year), which the valve is submitted to. Their effects are cumulative, and are expressed by lineal ruptures and/or perforations.

This means that it is difficult to determine the best moment for replacing the valve. The problem of the user getting it done too late must be balanced against the economic costs to the Social Security system, if is carried out unnecessarily early. So, a sensor is needed that can continuously monitor the working of the valve and give physicians objective data in order for them to make the substitution of the valve at the right time. This sensor has to be non-invasive and biocompatible. If small samples of soft magnetic material are inserted in the valve their movement can be detected when the valve is submitted to an alternating field of a frequency ranging from a few tens of Hz to tens of MHz, depending on the magnetic material used and its morphology (fig 7).

The material inside the body cannot interfere with the normal working of the valve. Each element's mass should not be higher than a few micrograms and its section must be of 10^{-8} m² or lower. Different soft magnetic materials meet these requirements as for example amorphous ribbons and wires made by melt spinning or amorphous microwires made by rapid quenching.

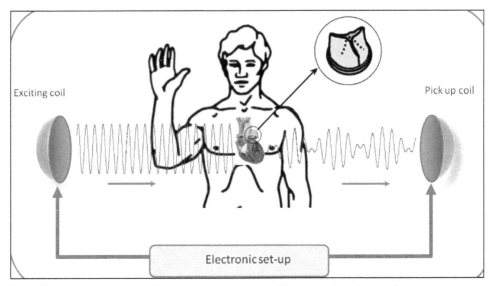

Fig. 7. Detection of motion of bioprostheses cusps with microwires inserted.

The best results are obtained by using an amorphous Co-based microwire as a magnetic element. This microwire has a diameter of 20 μm and is coated in the manufacturing process itself by a glass layer. The total diameter of the coated microwire is 45 μm. It presents remarkable advantages over other materials (Marin P. & Hernando A. 2004):

- It can be integrated in the cusps without the usage of biological glues.
- Due to its minimum size it neither diminishes the cusps flexibility, nor does it change their mass in an appreciable way.
- It is coated, during the process of manufacture, by a biocompatible glass, so facilitating its integration.
- Despite its reduced size the magnetic signal obtained is comparable to that obtained with the other models.
- Its high-shape anisotropy (10mmx45μm) together with its high-magnetic susceptibility makes it possibly to greatly reduce the energy expense needed for the sensor working.

The sensor works as follows. An excitation field is generated by an external coil. This field magnetizes the soft materials attached to the cusps and this magnetization triggers to a signal, the carrier signal, having the same frequency as that of the excitation one (2 kHz). The working of the valve causes a movement of the magnetic elements together with the cusps, producing a change in their position with respect to the exciting and detection coils (fig 7). This change in the relative position causes a modulation in the carrier signal (2 kHz) that clearly has the frequency of the opening and closing of the valve (cardiac rhythm, 60 – 80 cycles/min).

The dysfunctions produced in the movement of the cusps during the opening and closing of the valve, due to degradation, hardening or rupture processes, provoke a change in the modulated signal, which is monitored against the signal obtained in the standard working conditions. The comparison between both signals may allow the medical services to establish precisely the moment when the valve must be replaced (Rivero at al., 2007). All this may be done by wireless, without any need to implant, in the patient, any electrical or electronic device for detection or transmission, or to establish any type of material joint with the valve to the patient.

This sensor has been tested in a hydrodynamic setup. Figure 8 shows the signal that is received in the pickup coils. Three different series of images corresponding to the opening process of the valve with different damage levels can be seen. The electrical signal obtained by the sensor system is superimposed on the corresponding image. The first row (a) corresponds to the opening process of the undamaged valve. The second row (b) corresponds to the valve with a slightly glued commissure, and the third row (c) corresponds to the same valve but with the three commissures slightly glued. In the last case it can be seen that there is a dramatic change in the electrical signal, as well as in the movement of the cusps.

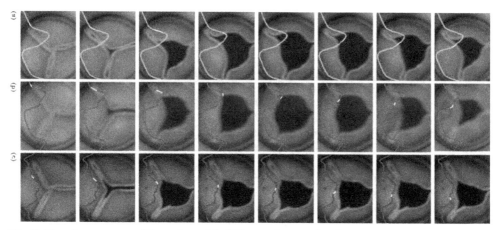

Fig. 8. Opening process of the valve with different damage levels.

3. Magnetic actuators

The working of most conventional electromagnetic actuators is based on the force exerted by a magnetic field on a moving element, made of a ferromagnetic element, that eventually can become a permanent magnet. The magnetic field that acts on the material can be created by a coil, with the appropriate feeding (with alternate or direct current) or by a permanent magnet. This force is used to change the moving element.

Other kind of magnetic actuators are based on the change in dimensions or the deformation that takes place in a magnetostrictive magnetic material (see section 2) when there exists a variation of the magnetic field that is acting on it. As in the previous case this field can be originated by a coil or a permanent magnet when it changes its position or its orientation. This kind of actuators, based on the effect commentated in section 2 of this chapter, present great advantages for their application on biomedicine since their magnetic field is not shielded by human body tissues due to their low conductivity. In other words the actuator field can be established by devices placed outside the human body although the actuator element is an internal implant. The limits imposed by the size and biocompatibility requirements are the same as the ones described in section 2 for the magnetic sensors.

There are other kinds of recent applications of magnetic actuators related to temperature. A general property of magnetic materials, except the ferrites, is an electrical conductivity in the range of metal and metal alloys, ranging between 10^4 and 10^7 S/m. This means that if the actuator field is an alternating one with a frequency ω, eddy currents will be induced in the element that will dissipate a Joule power proportional to the conductivity value, the square of the magnetic field amplitude and the square of the frequency.

$$W_{Joule} \propto \sigma \, \omega^2 \, B_0^{\,2} \tag{9}$$

Moreover, any ferromagnetic material will dissipate, through heat, the power needed for the magnetization process, which will be proportional to the frequency.

$$W_{Hyst} = \omega \int_{hystloop} H dB \tag{10}$$

If the frequency is high, (between 10 KHz and 10^3 KHz), the power dissipated through both causes may mean a significant local increase in the medium temperature. This behavior has been used by numerous researchers in the past decade in order to try to find a treatment against tumors (hyperthermia) which would be not too aggressive for the patient. In order to avoid the danger of creating an embolism in narrow capillaries, nanoparticles with diameters ranging from five to a few hundreds of nanometers are used. To avoid the particles aggregating because of the attraction of their magnetic moment, superparamagnetic materials or ferromagnetic materials functionalized with polar radicals that cause electrostatic repulsion are used. Another application of these particles, (in their hollow mode), is to use them as a vehicle for transporting medicines to the place in the organism where they have to act. Once they are there they are heated with a high frequency magnetic field that provokes their fragmentation and frees the medicine inside the nanoparticle.

These magnetic actuators have given rise to a new kind of materials, using traditional materials as a starting point but reducing them to smaller and smaller sizes. One only has to consider that in a gold cube with a 1 micron side just one part in a million of their atoms are arranged on the surface while in a cube of 2 nm 60 % of them are on the surface. So it is not strange that their physical properties become very different from that of bulk gold when reaching the limit where gold nanoparticles of diameter 1.5 nm demonstrate a spontaneous magnetic moment (Crespo et al., 2004).

3.1 Magnetic endoluminal artificial urinary sphincter

Urinary incontinence is not considered an illness but a symptom and is only treated when it becomes a social problem. Urinary incontinence is defined by the complaint of any involuntary leakage of urine (Abrams et al., 2002) or unintentional loss of urine that occurs with such frequency and in such quantities as to cause physical and/or emotional distress in the person experiencing it. Moreover, the number of people suffering from this is estimated to be around a million people among the adult population in western countries (Irwin et al., 2006).

Depending on the origin and the severity of the affection, it is treated with absorbent pads, pharmacological or surgical methods. For more than 30 years physicians have implanted inflatable artificial sphincters that simulate the sphincter function and permit the voluntary control of the micturition (American Medical System). However this device requires highly invasive surgery since it is necessary to implant a cuff around the urethra, in addition to a balloon, that regulates the cuff pressure and a bulb that controls the inflation and deflation of the cuff.

The objective of developing a magnetic artificial urinary sphincter is to look for a minimally invasive device that permits voluntary micturition control, making use of the potential of the magnetic field to exert a force without physical contact.

The preliminary specifications for the device are the following:

- Voluntary opening by the patient
- Automatic closing
- Automatic opening when the safety pressure is reached
- External magnetic drive, without other devices
- Urethral implantation without surgery
- Immunity to RF fields

The device consists of a magnetic valve placed in the urethra. The urine is evacuated by bringing a permanent magnet near the body of the patient. The magnetic valve consists on (Figure 9) a hollow cylindrical body valve with a toroidal magnet fixed at one of their ends and a soft magnetic material piston that is attracted by the magnet, closing the evacuation hole. As a sealing gasket a medical grade silicone O-ring is used.

If a more powerful external magnet is brought near the valve, the magnetization of the piston (made of a soft magnetic material) is reversed and a repulsion force between the internal magnet and the piston appears, causing the evacuation hole to open. When the external magnet is moved away, the interaction between the internal magnet and the piston turns attractive again and the valve closes automatically. Furthermore, the system is provided with a safety system preventing overpressure in the bladder. By adjusting the distance between the piston and the internal magnet, a pressure level can be fixed in such a way that if the pressure inside the bladder reaches that value, then the valve opens automatically.

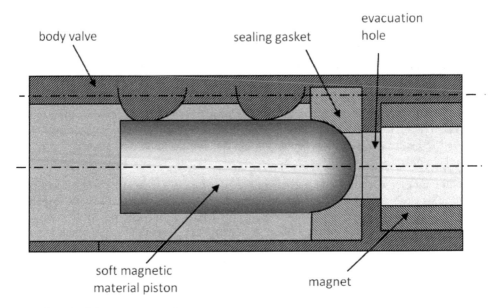

Fig. 9. Sketch of the magnetic valve.

Figure 10 shows the sphincter prototype for the trials performed on animal models, in the Service of Experimental Surgery of Puerta de Hierro Hospital.

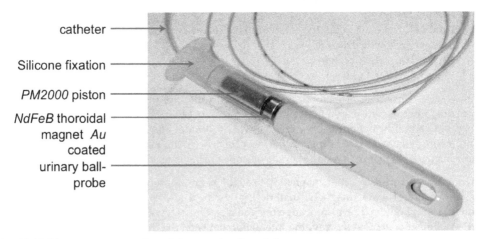

Fig. 10. Sphincter prototype for trials on animal models.

Fig 11 shows the operating system of the valve and its relative position concerning the bladder. Without any external action, the valve is in its closed position thanks to the action of the magnetic force exerted by the internal magnet (Fig11, A).

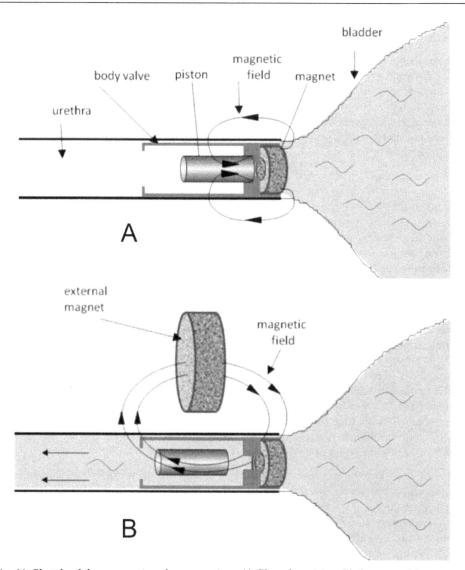

Fig. 11. Sketch of the magnetic valve operation. A) Closed position. B) Open position.

From the point of view of the biocompatibility of the system the ferromagnetic elements are the most critical ones. For the piston, a Fe based alloy is used, PM2000, that previous studies have proved, has a good biocompatibility (Flores et al., 2004) together with a good magnetization for the fields that the magnet produces at millimetres distances. Moreover, the NdFeB magnet has been coated with gold in order to isolate it from the biological medium. For the body valve, a medical grade polyurethane with biomedical grade has been chosen. The problem regarding the fixation to the urethra has been solved by adapting the valve to a urinary ball-probe. Trials on experimental animals have shown good preliminary results.

3.2 Hyperthermia HeLa cell treatment with silica-coated manganese oxide nanoparticles

The treatment of tumours by hyperthermia is based on the killing of tumour cells by heating them. Temperatures over around 42°C kill tumour cells. Moreover, the behaviour of normal and tumour cells is different with temperature; generally, normal cells show better resistance to temperature than the tumour ones (Hahn et al., 1974). If it is possible to heat the tumour area at temperatures that would kill tumour cells but not the normal cells, we would be able to treat tumours selectively with less damage to the body than other therapies such as chemotherapy.

As mentioned above, in the past years nanoparticles have attracted much attention for medical applications due to their small size that enables them to be inserted inside the body and transported round it. In particular, magnetic nanoparticles have been used as a heating source for magnetic hyperthermia. Under the influence of an alternating high-frequency magnetic field, they generate heat through hysteresis losses, induced eddy currents and Neel and/or Brown relaxation processes (Jordan et al., 1993). Thus, different kinds of magnetic nanoparticles have been tested as a heating source. The new materials must satisfy stringent conditions: they must be biocompatible, be stable in aqueous solution, possess high thermal efficiency as heating elements and be highly accumulating inside tumour cells, so that when applying the alternating magnetic field (AMF) the increase in temperature can induce cellular death (Kong et al., 2001). Superparamagnetic iron oxide is the most common material tested up to now due to its high biocompatibility, low synthesis cost, enhanced specific loss power and easy functionalization (Hergt et al., 1998). However, in spite of these advantages, it is not possible to control the local heating temperature, because it is not possible to measure the exact temperature or distribution of temperature in a tissue under magnetic particle hyperthermia (MPH) treatment. Therefore, the temperature reached in a tissue under MPH treatment will depend on a large number of particle parameters, such as size and concentration in the tissue of the nanoparticles, the conditions of the external applied field, and the length of the treatment (Gazeau et al., 2008).

To avoid the obstacle of temperature control, new materials of tunable Curie temperature (Tc) are being intensively investigated to achieve temperature autostabilization at the hyperthermia conditions (Pradhan et al., 2007). Magnetic particles with tuneable Tc will prevent the temperature of the whole tumour or the hottest spot around the particle raising its temperature over the Curie Temperature, so, avoiding the use of any local temperature control system.

When a ferromagnetic nanoparticle reaches its Curie Temperature it becomes paramagnetic, and its magnetic moment drastically decreases. Thus, the eddy currents and the relaxation processes decrease as well and the hysteresis losses disappear. This means that the nanoparticles will stop heating the medium and the temperature will go back below the Curie Temperature becoming ferromagnetic again and so recovering the heating process. Therefore, it becomes a self regulating system and the culture temperature will always be around the Curie Temperature.

Bearing in mind all these facts, manganese perovskites meet the requirements for magnetic hyperthermia treatments.

The manganese oxides perovskite $La_{1-x}(SrCa)_xMnO_3$ have a Curie temperature that, depending on the cation ratio, can range from 300 K to 350 K (so 42-44 °C is within the range) and they have large magnetization values of about 30 – 35 emu/g.

The preparation of the particles is as follows. First the particles are created by the ceramic method from compounds of La_2O_3, $CaCO_3$, $SrCO_3$, and MnO_2. These particles obtained by the ball milling method form agglomerates due to the dipolar magnetic interaction and the lack of surfactants. The agglomerates also have a large size distribution, with sizes greater than 1 μm. Since the magnetic interaction decreases with temperature, the particles are dispersed in ethanol and heated over the Tc in order to disaggregate them and to select the smaller ones, thus obtaining an average size of 100 nm. These NPs are not biocompatible so they have to be coated with silica following the Stöber method (Stöber et al., 1968).

The magnetic properties are not significantly affected by the size selection. Since the size is around 100 nm, the selected NPs still behave quite like the as-prepared ceramic. However, the magnetic properties are affected by the coating: the total magnetization is reduced by the presence of diamagnetic silica and the Curie Temperature decreases when the nanoparticles are coated. For example, for the composition $La_{0.56}(CaSr)_{0.22}MnO_3$, at low temperature the magnetization at 1 kOe decreases from 31 to 21 emu/g (about 32%) while the Tc decreases from 68 °C to 44 °C for the uncoated and the coated nanoparticles respectively.

Another problem that must be faced is if it is necessary to increase the temperature up to 42 – 44 °C in the whole tumour (so needing magnetic nanoparticles with very high magnetic moment or a very high concentration of nanoparticles) (Lacroix et al., 2009) in order to induce tumour damage or if it is enough to raise the temperature locally in the cells to induce apoptotic tumour death. The study of intracellular hyperthermia can shed light on this question.

Biological tests with perovskites were performed in order to prove their validity as a heating source for tumour hyperthermia. They were put inside a culture of HeLa cells. HeLa cells are a family of tumour cells widely used by biologists. HeLa cells were incubated with a concentration of 0.5 mg/ml of perovskites for 3 h. After incubation, cells were washed 3 times with PBS and then exposed for 30 min to a 100 KHz alternating magnetic field of 15 mT.

The cells incubated with the perovskites but without being submitted to an alternating magnetic field do not show any change, thus, demonstrating that the coating of the nanoparticles makes them biocompatible as expected. For the cells submitted to an alternating field the cell morphology was not affected immediately after the incubation and exposition to AMF. However, the perovskite + AMF treatment provoked deep morphological alterations, 24 h after the combined treatment, which corresponds to different stages of cell death by an apoptotic process. It is known that apoptosis is a regulated process which requires the active participation of specific molecules and is a characteristic mechanism of cell death for temperature around 42 °C. The temperature increase of the culture during the application of the AMF (controlled by an infrared thermometer) was lower than 0.5 K for the PER incubated HeLa cells. This means that the small size of the perovskites cannot heat the cell culture. However, perovskites can induce local hot spots that damage irreversibly the structure and functionality of the cell proteins triggering the cell apoptosis (Fig 12).

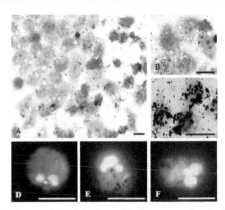

Fig. 12. Nanoparticles (black spots) inside the cell after apopthosis (White holes) (Villanueva, 2010)

These results show that perovskite nanoparticles have a high potential for cancer cell hyperthermia, working as smart mediators for self controlled heating of tumours, where the heating source is switched off when the local temperature of the tumour reaches the desired value.

4. Biocompatibility

In the previous paragraphs it has been shown that the different materials that are used as sensors should meet different requirements from the point of view of their magnetic properties: the optimal response of the device could be achieved with the softest, the hardest or the most magnetostrictive material. It depends on the application. It should be remarked that to assure a long-term working life-time of the device, it is necessary to look for a material with specific mechanical properties (González-Carrasco, 2009), as for example high fatigue resistance for the cardiac valve sensor. In the particular case of the implantable devices, it is evident that there is a decisive requirement that is more important than all those physical properties. This is, that the device cannot harm the patient, so the system must be biocompatible. The definition of the term Biocompatibility says: "Biocompatibility refers to the ability of a material to perform with an appropriate host response in a specific situation" (Williams, 1987).

Bearing in mind, that the human organism can be considered, in a very schematic way, as an assemblage of different tissues and fluids with different properties (biological, physical and chemical) as well as being a very aggressive medium against exogenous elements, it is easy to infer that the problem of the biocompatibility of the materials is very complex and should be studied for each specific application (Williams, 2008). In general, biocompatibility studies usually begin with in vitro tests (i.e. cell viability, preferably with cells of the host tissue/organ, and an endless quantity of trials concerned with the promotion or inhibition of biological process) followed by in vivo trials on experimental animals (Ratner et al., 2004).

Biomaterials, understood as materials that are biocompatible, can be classified in accordance with different parameters, for example their nature (metal, ceramic, polymer...), their application (bone, vascular, eye...) or their size. For reviewing the state of the biocompatibility of magnetic materials, this last option is more appropriate. Two large groups of materials can be distinguished: those used as bulk materials and the micro/nanoparticles.

On the one hand, among the widely established bulk biomaterials (titanium alloys, cobalt-chromium alloys, noble metals, Nitinol, austenitic stainless steel, alumina, calcium phosphates, carbon and polymers like UHMW polyethylene, PMMA or silicones) there are none with ferromagnetic properties. Regarding the metallic materials (excepting the noble metals, which are not ferromagnetic) only those that develop a well attached surface oxide layer usually present a good response to corrosion. In fact, if Chromium is added above 12% weight to Fe or Co the alloys become stainless. The stainless steels used for medical applications (316L and 304L) also incorporate Ni, among other elements, in their composition, which improves their corrosion resistance and stabilizes the face centred cubic structure. This latter fact explains their paramagnetic character.

In general, it could be said that most the ferromagnetic materials are not biocompatible or their biocompatibility is not known. Very few papers on the biocompatibility of magnetic materials can be found. Even fewer can be found on cells. Most of them are corrosion studies done with liquids that simulate the pH of different biological mediums. Magnets have attracted interest in medicine (Riley at al., 2002). In particular, rare earth magnets have been investigated in the field of dentistry to push or pull teeth or as prostheses retention systems (Noar et al., 1999). They show bad corrosion resistance and have been used embedded in different polymeric materials or with Ti coating.

Studies on Fe-Co alloys and magnetostrictive NiMnGa and Terfenol_D show poor cell viability, excepting the last one that presents high corrosion (Pouponneau et al. 2006). However the system Co-Pt and Fe-Pt show better corrosion response (Yiu et al. 2004)). At research level, the Fe based alloy PM2000 shows good corrosion behaviour and cell viability together with a significant saturation magnetization, especially when coated with alumina by thermal oxidation (Flores et al., 2004).

The surface is where the first contact between the material and the biological entity takes place. This is why surface biomaterials are being researched in depth, while surface modification is one of the most widely used strategies for improving biomaterial properties.

In the case where a magnetic feature can only be achieved with a material that is not biocompatible the simplest solution is to encapsulate it or to coat it with a biocompatible material. Anyway, as coatings also present their own problems (adherence, thickness,...), it would be very interesting to investigate and to develop biomaterials with good magnetic properties.

On the other hand, there already exist some nanoparticles commercially used as contrast agents in imaging diagnostic techniques or drug targeting and magnetic separation applications, like the Iron Oxide or the Gd, because of their magnetic properties. However, an enormous effort is being made to develop biocompatible magnetic nanoparticles for their application in biomedicine due to the attractive possibilities that they offer (Pankhurst, 2003) as hyperthermia agents for coadyuvant cancer treatment, drug delivery systems, as well as for the previously mentioned reasons.

The problem of the toxicity of nanoparticles goes further than in vitro trials with cells, (it can be really complex) (Lewinsky, 2008). In fact this kind of material has other problems when used for long term devices. For example, once their behaviour with cells has been tested, it is essential to understand their survival in the body. Do the macrophages detect them too soon or on the contrary, do the nanoparticles tend to accumulate in some important organ such as the liver or brain? (Hoet et al., 2004). Nanoparticles are frequently coated with different organic or inorganic materials (as dextran (Lacava, 2001) or silica (Villanueva, 2010)). Sometimes this is done to make them biocompatible, sometimes to obtain an appropriate dispersion and sometimes to functionalize them.

5. Further research

Magnetic sensors can be applied in several fields of Biology and Medicine. Practically all the devices mentioned in this chapter are still in the optimization phase or in clinical trials. So, the related research is still ongoing.

The hyperthermia treatment studies for malignant tumors merit, because of their enormous interest, to be discussed separately. This is research with a marked interdisciplinary character, since it presents challenges that specifically concern doctors, biologist, chemists, material´s engineers and, of course, physicists. Although nowadays, nanoparticles have been achieved that seem to work very well, there remain problems that are far from being solved.

- To determine if the origin of the damage produced to the tumor cells comes from the temperature increase or from the movement of the particles inside the cells themselves.
- To concentrate the particles in the nearby tumor area and avoiding most of them being eliminated by the reticulum-endothelial along the path to the tumor.
- To control the temperature treatment inside a medium that the circulatory system keeps constantly at 37ºC.
- To eliminate the particles from the patient´s body after the treatment and avoiding their concentration in organs, such as the kidneys, liver, spleen, etc.
- To concentrate the high frequency magnetic field in the tumor area, in order to avoid heating the particles that have been absorbed by other organs or tissues.
- Last but not least, to manufacture cheap nanoparticles so that if the treatment works, everybody can have access to it.

Another area of research we are interested in is the application of magnetic sensors and actuators in remote monitoring of multiple sclerosis patients. Multiple sclerosis (MS) is a chronic, degenerative disease that affects young people and which , according to the last survival studies, not only shortens life approximately by 10 years but also produces important limitations and deficiencies affecting the wellbeing of the patient, over the course of its development.

During its progress, MS patients present a very varied symptomatology. So, in the first stages of the disease it is usually asymthomatic and without objective neurological deficiencies. It is only during the 5 first years that objective neurological manifestations begin to appear. As a general rule, after between 10 to 15 years of development, the deficiencies are very evident and begin to have clear repercussions on the physical and cognitive activities of the patients. Between the 15th and 20th years of development, the limitations increase by up to 33%, which means severe restrictions on mobility. In these advanced stages of the disease, the symptoms limit the patient's mobility very severely and drastically affect the quality of life and autonomy of these patients.

The goal of this research would be to exploit the possibilities offered by telemedicine and remote control for improving the assistance, monitoring and quality of life in MS patients, in the advanced stages and with very limited mobility.

It would be focused on three aspects:

- Symptoms monitoring:
    - Spasticity
    - Dysphagia
    - Pain
    - Autonomic alteration: cardiac rhythm, blood-pressure variations, gastrointestinal mobility alterations and sphincters control (anal and urethral)
    - Position
    - Sleep quality
- Development of complications derived from the physical condition
    - Decubitus ulcers
    - Urine infection
    - Respiratory infection (aspiration)
    - Position alteration (orthopaedic malformation)
- Clinical evolution monitoring (search of image subrogated markers obtained by periodical and quantifiable Magnetic Resonance studies)
    - Walk monitoring
    - Upper limbs monitoring
    - Mental function monitoring
    - Autonomic state monitoring

These neurological requirements scenarios could be analyzed at a distance by using the appropriate sensors (with calls and alerts), so that the neurologist would be able to treat the patient without the patient needing to be in presence of the physician and so avoiding complications that would decrease the quality of life.

6. Conclusion

It is not possible in the space available to offer an exhaustive overview of the applications of magnetic sensors in the field of medicine. We have limited ourselves to presenting just some particular works, that we have recently developed, some that are currently being developed and some that are still in the future.

It seems obvious that the development of magnetic sensors and actuators is generally linked to the development of magnetic materials. Furthermore, it could be said that the development of our technological civilization is linked to the development of magnetic materials. If this assertion sounds too exaggerated, please imagine the consequences of the disappearance of the silicon steel sheet and with it all the electric motors and transformers. Or just think about the permanent magnets inside all mobile phones, or the magnetic information storage medium. Or, going back in time, at what stage of development would we be without the great discoveries made possible by the compass.

This is even more evident in the case of the biomedical applications. The increasingly strict requirements for sensors and actuators in the field of safety, as well as for biocompatibility, corrosion, miniaturization, low consumption, etc. are impossible to meet without new magnetic materials. Just think that the fight against the most feared disease of our time could depend on something so simple (but not easy) as the suitable magnetic material.

7. Acknowledgment

The authors wish to acknowledge the grants provided by Ministerio de Educación of Spain, Comunidad de Madrid, Ministerio de Sanidad of Spain, Laboratorios Indas S. A., Fundación Mutua Madrileña and Molecular Nanoscience Consolider-Ingenio CSD2007-0010-(2) for the realization of the works presented in this chapter.

Also they want acknowledge the collaboration of the electronic enterprise Advanced Processing Machines S. L.

8. References

Abrams P, Cardozo L, Fall M, Griffiths D, Rosier P, Ulmsten U, van Kerrebroeck P, Victor A & Wein A. (2002). The standardization of terminology in lower urinary tract function: report from the standardization sub-committee of the International Continence Society. *Neurology and Urodinamics*. Vol. 21, pp. 167-178.

Askew F, Broughall JM, Griffiths J, Hyland M & Lorimer K. Electrochemical sensor. Patent ES2345938. Spain (06.10.2010).

Cai, QY, Jain, M K, Grimes CA. (2001). A wireless, remote query ammonia sensor. *Sensors and Actuators* B 77, pp. 614–619.

Committee for the National Institute for the Environment, 2005. Congressional Research Service, http://www.cnie.org/nle/AgGlossary/letter-s.html.

Crespo P, Litrán R, Rojas T C, Multigner M, de la Fuente J M, Sánchez-López J C, García M A, Hernando A, Penadés S and Fernández A (2004), Permanent magnetism, magnetic anisotropy and hysteresis of thiol-capped gold nanoparticles. *Physical Review Letters*. Vol 93, pp. 087204-1-087204-4.

Drobrovol´skii NA, Kostriso PR, Labinskaya TA, Makarov VV, Parfenov AS & Peshkov AV. (1999) . A blood coagulation Analyzer. *Biomedical Engineering*. Vol. 33, pp. 44-47.

Flores MS, Caipetti G. González-Carrasco JL, Montealegre MA, Multigner M, Pagani S & Rivero G. (2004). Evaluation of magnetic behaviour and in vitro biocompatibility of ferritic PM2000 alloy. *Journal of Materials Science: Material in medicine*. Vol. 15, pp. 559-565

Gazeau F, Levy M & Wilhelm C.(2008). Optimizing magnetic nanoparticle design for nanothermotherapy. *Nanomedicine*. Vol. 3, pp.831-844.

González-Carrasco JL, (2009). Metals as bone repair material. Cap. VI, *In: Bone repair biomaterials*, Ed: JA Planell, SM Best, D. Lacroix andA. Merolli. CRC Press and Woodhead Publishing Limited, ISBN: 978-1-84569-385-5.

Grimes A, Ong KG, Loiselle K, Stoyanov PG, Kouzoudis D, Lin Y, Tong C &Tefikn F. (1999). Magnetoelastic sensor for remote query environmental monitoring. *Smart Material an Structures*. Vol. 8, pp.639-646.

Hahn GM. (1974). Metabolic aspects of role of hyperthermia in mammalian cell inactivation and their possible relevance to cancer treatment. *Cancer research*. Vol. 34, pp. 3117-3123.

Hergt R, Andra W, d'Ambly CG, Hilger I, Kaiser WA, Richter U & Schmidt HG.(1998). Physical limits of hyperthermia using magnetite fine particles. *IEEE Transations on Magnetism*. Vol. 34, pp 3745-3754.

Hernando A, García-Escorial A, Ascasibar E &Vázquez M. (1983). Changes in the remanent magnetisation, magnetoelastic coupling and Young's modulus during structural relaxation of an amorphous ribbon. *Journal of Physics D: Applied Physics.* Vol.   16, pp.1999-2010.

Hoet P.H.M., Brüske-Hohlfeld I. & Salata O.V. (2004). Nanoparticles-known and unknown health risks. *Journal of nanobiotechnology*, Vol. 2, 12.

Irwin D.E., Milsom I., Hunskaar S., Reilly K.,Kopp Z.,Herschom S., Coyne K.,Kelleher C., Hampel C., Artibani W.& Abrams P. (2006). Population-Based Survey of Urinary Incontinence, Overactive Bladder, and Other Lower Urinary Tract Symptoms in Five Countries: Results of the EPIC Study. *European Urology.*Vol. 50; pp. 1306-1315.

Jordan A, Scholz R, Wust P, Fähling H & Felix R. (1999). Magnetic fluid hyperthermia (MFH): Cancer treatment with AC magnetic field induced excitation of biocompatible superparamagnetic nanoparticles. *Journal of Magnetism and Magnetic Materials*. Vol. 201, pp. 413-419.

Kong G, Braun RD, Dewhirst MW.2001. Characterization of the effect of hyperthermia on nanoparticle extravasation from tumour vasculature. *Cancer Research*. Vol. 61, pp. 3027-3032.

Kouchoukos NT, Blackstone EH., Doty DB, Hanley FL &Karp RB (2003). *Kirklin/Barratt-Boyes Cardiac Surgery, 3rd ed.,* Churchil Livingstone, ISBN: 978-0-443-07526-1,Philadelphia, 2003, pp. 554–656.

Lacava LM, Lacava ZGM, Da Silva MF, Silva O, Chaves SB,Azevedo RB, Pelegrini F, Gansau C, Buske N, Sabolovic D & Morais PC. (2001). Magnetic resonance of a dextran-coated magnetic fluid intravenously administered in mice. *Biophysics Journal*. Vol. 80, pp.2483–2486.

Lacroix LM, Malaki RB, Carrey J, Lachaize S, Respaud M, Goya GF & Chaudret B. (2009). Magnetic hyperthermia in single-domain monodisperse FeCo nanoparticles: Evidences for Stoner-Wohlfarth behaviour and large losses. *Journal of Applied Physics*. Vol. 105, pp. 023911 1-4.

Lakshmanan R, Guntupalli R, Hu J, Petrnko VA, Barbaree JM,& Chin BA.(2007). Detection of Salmonella typhimurium in fat free milk using a phage immobilized magnetoelastic sensor. *Sensors and Actuators B*. Vol. 126, pp.544-550.

Lewinski N, Colvin V. & Drezek R. (2008). Cytotoxixity of nanoparticles. *Small*. Vol. 4, pp.26-49.

Marín P & Hernando A. (2004). K.H.J. Buschow, R.W. Cahn, M.C. Flemings, B. Ilschner, E.J. Kramer, S. Mahajan, P. Veyssière (Eds.), Magnetic Microwires: Manufacture, Properties and Applications.*In:. Encyclopedia of Materials: Science and Technology*, Elsevier. ISBN:0-08-043152-6.Oxford.

Neel GT, Parker JR, Collins RL, Storvick DE, Thomeczek CL, Murphy WJ, Lennert GR, Young MJ & Kennedy DL.(1998). Fluid dose, flow and coagulation sensor for medical instrument. Patent US5789664. United States (04.08.1998)

Noar JH & Evans RD. (1999). Rare earth magnets in Orthodontics: an overview. *British Journal of Orthodontics*, Vol. 26 , pp. 29-37.

Pankhurst QA, Connolly J, Jones SK & Dobson J.(2003). Applications of magnetic nanoparticles in biomedicine. *Journal of Physics D: Applied Physiscs.* Vol. 36, pp. R167-R181.

Pouponneau P., Yahaia L.H., Merhi Y., Epure L.M. & Martel S.(2006). Biocompatibility of candidate materials for the realization of medical devices. *Proceedings of the 28th IEEE EMBS Annual International Conference,* ISBN: 1-4244-0033-3/06, New York City, Aug-Sept,2006.

Pradhan P, Giri J, Samanta G, Sarma HD, Mishra KP, Bellare J, Banerjee R, Bahadur D.(2007). Comparative evaluation of heating ability and biocompatibility of different ferrite-based magnetic fluids fro hyperthermia application. *Journal of biomedical materials research Part B: Applied Biomaterials.*Vol. 81B, pp-12-22.

Ratner BD, A S Hoffman, Schoen FJ & Lemons J. (2004) Biomaterials Science. *An Introduction to Materials in Medicine (2nd edition).* Academic Press. ISBN: 978-0125824637. San Diego.

Riley MA, Walmsley AD, Speight JD & Harris IR. (2002). Magnets in Medicine. *Materials Science and Technology* . Vol. 18, pp.1-11.

Rivero G , Crespo P, Spottorno J, de la Presa P, Multigner M, Valdés J, Villanueva Mª A, Cañete M & Morales MP. "Sensor system for continuously monitoring the cellular growth "in situ" based in magnetoelastic sensor", Patent P200801973 Spain (01.07.2008).

Rivero G, García-Páez JM, Álvarez L, Multigner M, Valdés J, Carabias I, Spottorno J & Hernando A. (2007) Magnetic sensor for early detection of heart valve bioprostheses failue. *Sensors Letters.* Vol 5, pp.263-266.

Rivero G, García-Páez JM, Álvarez L, Multigner M, Valdés J, Carabias I, Spottorno J & Hernando A. (2008). *Sensor and Actuators A: Physical.* Vol. 142, pp. 511-519.

Vázquez M. & Hernando A., (1996). A soft magnetic wire for sensor applications. *Journal of Physics D: Applied Physics* Vol. 29, pp. 939-949.

Villanueva A, de la Presa P, Alonso J M, Rueda T, Martínez A, Crespo P, Morales M P, Gonzalez-Fernandez M A, Valdés J, Rivero G. (2010) *Journal of Physical Chemistry C.* Vol. 114, pp. 1976-1981.

Shen W, Lakshmanan R S, Mathison LC, Petrenko VA & Chin BA. (2009). Phage coated magnetoelastic micro-biosensors for real-time detection of Bacillus anthracis spores. *Sensors and Actuators B: Chemical* Vol. 137, pp. 501-506.

Wan J, Shu H, Huang S, Fiebor B, Chen IH, Petrenko VA. (2007). Phage-based magnetoelastic wireless biosensors for detecting Bacillus anthracis spores. *IEEE Sensors Journal.* Vol. 7, pp. 470-477.

Williams DF. (1987). *Definitions in biomaterials.* Elsevier. ISBN: 9780444428585. Amsterdam.

Williams DF. (2008). On the mechanisms of biocompatibility. *Biomaterials.* Vol. 29, pp. 2941-2953.

Xie F, Yang H, Li S, Shen W, Johnson ML, Wikle HC, Kim DJ, Chin BA. (2009). Amorphous magnetoelastic sensors for the detection of biological agents. *Intermetalics.* Vol. 17, pp. 270-273.

Yiu EYL, Fang DTS, Chu FCS &Chow TW. (2004) Corrosion resistance of iron-platinum magnets. *Journal of dentistry* . Vol. 32, pp. 423-429.

7

Magnetic Field Sensors Based on Microelectromechanical Systems (MEMS) Technology

Maria Teresa Todaro[1], Leonardo Sileo[1,2] and Massimo De Vittorio[1,2,3]
[1]National Nanotechnology Laboratory (NNL), Istituto Nanoscienze-CNR
[2]Center for Biomolecular Nanotechnologies UNILE, Istituto Italiano di Tecnologia
[3]University of Salento, Lecce
Italy

1. Introduction

In the last decades magnetic field sensors have been developed and realized for analyzing and controlling thousands of functions (Ripka, 2001), and they have become a widespread presence in modern lifestyle. Numerous applications in different fields of science, engineering, and industry rely on the performance, ruggedness, and reliability of magnetic field sensors.

The applications of magnetic sensors depend on magnetic field dynamic range and resolution and include position sensing, speed detection, current detection, non-contact switching, space exploration, vehicle detection, electronic compasses, geophysical prospecting, non-distructive testing, brain function mapping (Lenz & Edelstein, 2006).

Nowdays there is an increasing requirement for magnetic devices with improved sensitivity and resolution, trying to keep as low as possible their cost and power consumption. Additionally there is the need to develop compact devices with several sensors able to measure different parameters including magnetic field, pressure, temperature, acceleration. In this way a multifunctional device could be integrated on the same substrate containing transducers and electronic circuits in a compact configuration without affecting device performances.

In this context microelectromechanical systems (MEMS) technologies play a prominent role for the development of a new class of magnetic sensors.

In general MEMS devices are miniaturized mechanical systems produced using fabrication techniques already explored in the electronics industry. The exploitation of MEMS technology for device fabrication not only makes possible the reduction of the device dimensions on the order of micrometers, but also allows the integration of the mechanical and electronic components on a single chip. In addition to the small device size this involves other important advantages such as light weight, minimum power consumption, low cost, better sensitivity and high resolution. This technology was successfully employed for the realization of portable devices such as gyroscopes (Chang et al., 2008), accelerometers (Li et al., 2011), micromirrors (Singh et al., 2008), and pressure sensors (Mian & Law, 2010).

Magnetic field sensors based on MEMS technology, depending on their operation principle and magnetic range, have a great potential for numerous applications in several fields spanning from vehicle detection and control to mineral prospecting and metal detection as well as to non-distructive testing and medical diagnostics.

This paper aims at the description of current research status in magnetic field sensors focusing on devices fabricated by exploiting MEMS technologies. The paper presents advances in the classes of devices that take advantage from these technologies to scale down magnetic sensors size, namely resonant sensors, fluxgate sensors and Hall sensors.

Resonant sensors exploit Lorentz force principle on micromachined structures excited at one of their resonating modes. These sensors can detect magnetic fields with sensitivity up to 1 T and a maximum achievable resolution of 1 nT.

Fluxgate sensors are inductively working sensors consisting of excitation and sensing coils around a ferromagnetic core. Such sensors can detect static and low frequency magnetic fields up to approximately 1 mT with a maximum resolution of 100 pT.

Hall sensors are based on Hall effect transduction principle and measure either constant or varying magnetic field. They have a magnetic field sensitivity range from $1\mu T$ to $1T$.

Following the introduction, the paper is organized as follows. The second section, is devoted to the resonant sensors, including the Lorentz force operation principle, examples of realized devices reported in the literature with an highlight on the employed technologies for the fabrication. Third section is focused on fluxgate microsensors including operation principle, state of the art and involved fabrication technologies. Fourth section is dedicated to the description of the Hall effect and Hall magnetic sensors employing MEMS technologies are reported. The fifth section describes the possible applications of this new class of compact devices. Finally in the section sixth the paper ends with the conclusion.

2. Resonant magnetic sensors

Resonant sensors exploit Lorentz force of resonating micromachined structures. These sensors can detect magnetic fields up 1 mT with a resolution down to 1 nT. Such devices are normally based on MEMS technologies, are small in size (order of millimetres), and promise all the advantages related to the employment of fabrication microtechnologies including multifuntionalities and integration of mechanical and electronic components on a single chip.

Resonant magnetic field sensors use resonant structures that are excited at their resonant frequencies by Lorentz forces. Such devices are able to give an amplified response if excited at frequencies equal to the resonant frequencies or vibrational modes of the structures (Bahreyni, 2008).

These structures commonly consist of clamped-free beams or clamped-clamped beams or torsion/flexion plates. In figure 1 it is shown a schematic diagram of the Lorentz force principle acting on a clamped-clamped beam resonant structure. This device can be designed for example to resonate to its first resonant frequency, associated to its first flexural vibration mode. This beam, exposed to an excitation source with a frequency equal to its first resonant frequency, will have a maximum deflection at its midpoint. In order to excite the device a metallic loop is placed on the clamped-clamped beam surface where an excitation current (I) flows inside it with a frequency equal to the first resonance frequency. When the beam is exposed to an external magnetic field (B_x) in the x-direction, then a Lorentz force (F_L) is generated.

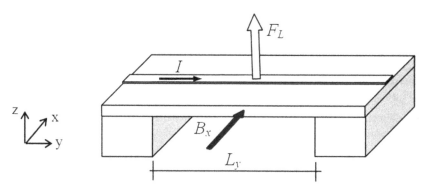

Fig. 1. Schematic diagram of the Lorentz force principle acting on a clamped-clamped beam.

This force can be determined as:

$$F_L = IB_x L_y \tag{1}$$

where the flowing current can be expressed as:

$$I = \sqrt{2}I_{rms} \sin(2\pi ft) \tag{2}$$

where L_y is the length of the metallic loop perpendicular to the magnetic field, I_{rms} is the root mean square of the current I, f is the frequency and t is the time.

The Lorentz force acts as an excitation source on the clamped-clamped beam, causing an amplified deflection on the midpoint. Thus, the magnitude of the beam deflection depends on the Lorentz force amplitude, which is directly proportional to I and B_x.

The application of an external magnetic field alters deflection/torsion of resonating structures with different shapes that is detected by exploiting different readout techniques.

In fact such deflections/torsions result in strain which is related to the elastic modulus of the structure material, to the geometrical characteristics of the resonating structures and to the quality factor (Herrera-May et al. 2010).

The quality factor is an important parameter of the resonant structures. It defines the bandwidth of the resonator relatively to its central resonant frequency or equivalently it expresses the maximum amplitude of the bending structure taking into account the different damping sources (Elwenspoek & Wiegerink, 2001, Beeby et al., 2004). High quality factors involve better device performance, better resolution and improved insensitivity to the disturbances (Beeby et al., 2004).

Another parameter of interest in resonant structures is the resonance frequency. Its determination can be obtained by using both analytical models and simulation tools and depends on elastic modulus, density, deflection and geometrical features of the resonant structure.

Moreover resonance frequencies are affected by residual stresses on the structure (Weaver et al. 1990). For example thermal stresses inside resonant devices (Hull, 1999) causes strains in the structures which in turn involve (Sabaté et al., 2007) a shift of the resonant frequency of the structures.

Such sensors are typically fabricated in silicon and polysilicon and main disadvantage of this technology is the resonance frequency shift due to temperature changes and environmental pressure which requires compensation electronic circuits and packaging under vacuum respectively.

To detect deflection of resonant structures different readout techniques have been used including the employment of piezoresistive, optical or capacitive techniques.

Piezoresistive sensing exploits changes in the resistance of piezoresistive elements placed in the hinges of the resonant structure, to detect changes in the output voltage signal as effect of strains originating from motions of beams or plates due to the Lorentz force.

Herrera-May et al. (2009) reported on a magnetic field microsensor based on a silicon resonant microplate ($400 \times 150 \times 15 \ \mu m^3$) and four bending microbeams ($130 \times 12 \times 15 \ \mu m^3$).

Figure 2 shows a schematic diagram of the fabrication process of the device.

The fabrication process is based on bulk micromachining technology on (100) 4" silicon-on-insulator (SOI) wafers. The process starts by growing a thin thermal oxide layer and depositing a silicon nitride layer on a SOI n-type substrate. The nitride layer is removed from the front side of the wafer and is patterned on the backside (figure 2(a)). Using a second mask, boron is implanted to create four p-type piezoresistors (figure 2(b)). A 1 μm-thick oxide layer is then grown and patterned. The area contacts ($120 \times 120 \ \mu m^2$) are opened (figure 3(c)) and then an aluminum layer is deposited and patterned to define metallic lines and pads (figure 4(d)). At this time the silicon substrate is etched from the backside using KOH that stops at the SOI buried oxide (figure 2(e)), which is then removed.

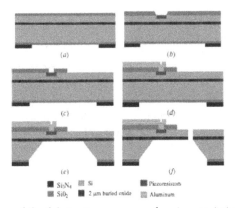

Fig. 2. Schematic diagram of the fabrication process of a piezoresistive resonant magnetic sensor reported by Herrera-May et al. (2009)

Finally, the SOI layer is etched by reactive ion etching (figure 2(f)) to define the plate-beam structure.

Figure 3 shows a schematic design of the resonant magnetic field microsensor reported by Herrera-May et al. (2009) with an highlight on the plate-beam structure and its working principle.

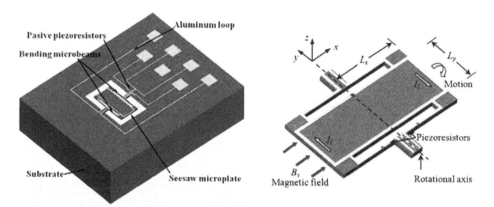

Fig. 3. Schematic design of resonant magnetic field microsensor (left) and highlight on the plate-beam structure and its working principle (right) reported by Herrera-May et al. (2009).

One of the main elements of this sensor is the aluminum rectangular loop deposited on the silicon plate. The Lorentz force causes a seesaw motion on the microplate and the bending of microbeams. Four piezoresistors (p-type) are connected in a Wheatstone bridge and two of these are active piezoresistors located on the microbeams. The Lorentz force originates a longitudinal strain on the two active piezoresistors changing their resistance. The change in the resistance of the active piezoresistors produces an output voltage shift of the Wheatstone bridge.

This sensor has a resonant frequency of 136.52 kHz, a quality factor of 842 at ambient pressure, a sensitivity of 0.403 $\mu V \mu T^{-1}$, a resolution of 143 nT with a frequency variation of 1 Hz, and power consumption below 10 mW. However, the sensor registered an offset and linearity problems in the low magnetic field range.

Tapia et al. (2011) reported on a piezoresistive resonant magnetic microsensor with seesaw rectangular loop of beams reinforced with transversal and longitudinal beams. This device was designed to be compact and to have high resolution for neurobiological applications. Characteristics of this microsensor are a resonant frequency of 13.87 kHz with a quality factor of 93, a resolution of 80 nT, a sensitivity of 1.2 VT^{-1} and a power consumption of 2.05 mW at ambient pressure. This sensor requires a simple signal processing circuit.

Other examples of piezoresistive magnetic sensors on the microscale have been reported in the literature (Beroulle et al. (2003), Sunier et al. (2006)).

Among the resonant magnetic sensors, there are some of them exploiting the optical detection.

Keplinger et al. (2004) reported a resonant magnetic field sensor using U-shaped silicon microbeams and an optical readout system. Figure 4 shows a schematic sketch of the device reported by these authors.

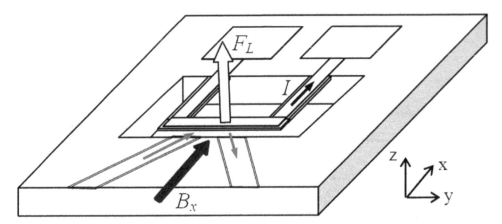

Fig. 4. Schematic sketch of the resonant magnetic sensor with optical readout reported by Keplinger et al. (2004).

Devices are processed by both surface micromachining and bulk micromachining techniques. The etching process has been also used to form the groove for the optical fibers. The microbeams contain a gold loop with a thickness of 500 nm. A magnetic field and an ac electrical current generate a Lorentz force, which bends the microbeams. These deflections are measured by optical sensing using an arrangement consisting of two fibers to avoid interference of reflected light. Figure 4 shows a design, in which the emitted light beam is reflected only once at the microbeam front side. This sensor can measure magnetic fields from 10 mT to 50 T at moderate excitation amplitudes. It can be used in harsh environments under mechanical vibrations and low temperatures. The device has a resonant frequency around 5 kHz, a resolution of 10 mT, and a power consumption of few milliwatts. The device needs high current magnitudes thus increasing the temperature and deformation at the silicon microbeam, with a possible resonant frequency shift.

Another example of magnetic sensor based on MEMS technology and having an optical readout has been reported by Wickenden et al. (2003).

Ren et al. (2009) reported on a resonant device exploiting the capacitive readout. The magnetic field sensor has been fabricated by using conventional MEMS technology and silicon-to-glass anodic bonding process. The device consists of a low-resistivity silicon plate suspended over a glass substrate by two torsional beams, as shown in figure 5. This silicon plate acts as electrode of sensing capacitances. Au capacitance plates are fabricated on the glass substrate and a multi-turn coil (Cr and Au layers) is deposited on silicon-plate surface. The Lorentz force causes an oscillating motion of silicon plate around the torsional beams, which produces a capacitance shift between the Au electrodes and the silicon plate. A capacitance detection circuit measured the capacitance change that depends on the magnitude and the direction of the external magnetic field. This sensor required a vacuum packaging to increase its performance. For a pressure of 10 Pa and 150 mV driving voltage amplitude, the microsensor has a resolution of 30 nT in the linear range from 3 µT to 30 µT, a sensitivity of 481 mVT^{-1}, a resonant frequency close to 1380 Hz, and a quality factor around 2500. Nevertheless, it presented a non-linear response from 0 to 3 µT.

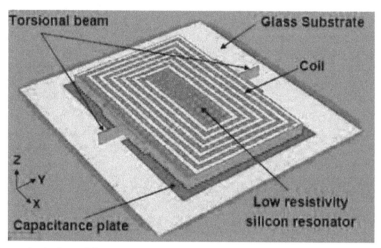

Fig. 5. Schematic design of the resonant magnetic sensor exploiting the capacitive sensing and reported by Ren et al. (2009).

Brugger et al. (2009) reported on a complex magnetic field sensor with a size of 7.5 mm × 3.2 mm, consisting of an electrostatically driven silicon resonator characterized by interdigitated combs for electrostatic excitation and capacitive detection, an amorphous magnetic concentrator and a pair of planar coils. It requires a complex fabrication process combining MEMS technology based on a silicon-on-insulator (SOI) substrate, the epoxy-resin-based attachment of a thin amorphous magnetic ribbon structured by wet chemical etching, micropatterning of the magnetic concentrator by UV-laser and vacuum packaging.. For a coil current of ±120 mA, the device offers a sensitivity of 1.91 MHzT^{-1} and a resolution of 1.3 µT. Under a pressure of 10^{-5} mbar, this microsensor presents a sensitivity of 1 MHzT^{-1}, a resolution of 400 nT, and a quality factor around 2400. It does not need a complex feedback and modulation electronics.

J. Kyynäräinen et .al. (2008) reported on resonant micromechanical magnetometers based on capacitive detection for 3D electronic compasses. The sensors has been fabricated by exploiting aligned direct bonding of a double side polished silicon wafer and a SOI wafer. Devices operated in vacuum to reach high enough Q values. Magnetometers measuring the field component along the chip surface have a flux density resolution of about 10 nT/$\sqrt{\text{Hz}}$ at a coil current of 100 µA. Magnetometers measuring the field component perpendicular to the chip surface are currently less sensitive with a flux density resolution of about 70 nT/$\sqrt{\text{Hz}}$.

There are in the literature other works reporting on capacitive sensing-based resonant magnetic sensors (Emmerich et al. (2000), Kádár et al. (1998), Tucker et al. (2000)).

3. Fluxgate sensors

Fluxgate sensors are inductively working sensors composed of excitation and sensing coils around a ferromagnetic core for detecting static and low frequency fields.

Fluxgate sensors can detect magnetic fields up to approximately 1 mT with a maximum resolution of 100 pT. Classical fluxgate sensors are expensive and have a big size. However in recent years a great effort was devoted to manufacturing micro fluxgate sensors using microfabrication technologies. Beside the small size, the advantages of micro fluxgate sensors are small weight, low power consumption, low cost in mass production and the possibility of on-chip electronics integration. The principal disadvantage is related to the fluxgate sensor parameters dramatically degrading when reducing the core size, which leads to low sensitivity and high noise level.

The fluxgate operation principle can be illustrated with the simple layout in figure 6 (Ripka, 2001). A ferromagnetic core immersed in an external magnetic field B_x is surrounded by an excitation coil which provides an ac excitation current I. This current periodically saturates the soft magnetic material of the sensor core at a frequency twice the excitation frequency.

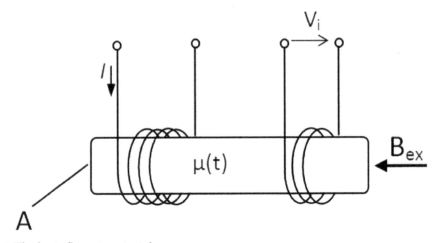

Fig. 6. The basic fluxgate principle.

A voltage in the pick-up coil (V_i) is generated due to the changing magnetic flux (Φ). From the Faraday's law:

$$V_i = \frac{d\Phi}{dt} = \frac{d(NA\mu_0\mu_r(t)H(t))}{dt} \qquad (3)$$

where μr is the relative permeability, μ_0 is the vacuum permeability, N is the number of turns and A is the cross sectional area, that we consider here as a constant.

By expanding, the equation (3) becomes:

$$V_i = \frac{NA\mu_0\mu_r dH(t)}{dt} + \frac{NA\mu_0 H d\mu_r(t)}{dt} \qquad (4)$$

where H is the field in the core material and is lower than the measured field H_{ex} in the open air due to demagnetization (Bozorth & Chapin (1942)):

$$H = H_{ex} - DM \tag{5}$$

where D is the demagnetising factor, $H_{ex} = B_x / \mu_0$ and M is the magnetization.

The first term in the equation (4) is the basic induction effect, and causes interference. Fluxgate operation is based on the second term, due to the variation of the core permeability with the excitation field. By considering the effect of demagnetization, the basic fluxgate equation becomes (Primdahl, 1979):

$$V_i = NA\mu_0 H_{ex} \frac{1-D}{(1+D(\mu_r(t)-1))^2} \frac{d\mu_r(t)}{dt} \tag{6}$$

The output voltage is on the second harmonics of the excitation frequency, as permeability reaches its minimum and maximum twice in each excitation cycle.

In accordance with the shape of the magnetic core, parallel-type fluxgate sensors fall into the categories of single core, dual core, ring-type core, racetrack type core (Ripka, 2001). The configuration of figure 6 is single core type. In order to eliminate the induction effect, a dual core configuration has been proposed, as showed in figure 7.

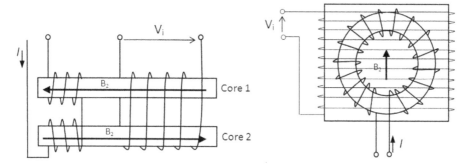

Fig. 7. Dual core (left) and ring type (right) configurations of a fluxgate sensor.

The driving coil is wound in opposite direction around the two cores, thus the induced magnetization fields are opposite in sign. If no external field is applied, the voltage induced in the sensing coil is zero in the ideal case. When an external field is present, a voltage is induced due to the differential change of the permeability (Primdahl, 1979).

High sensitivity can be achieved by increasing the number of turns N (if N is very high coil parasitic capacitance limits the sensitivity), by decreasing the demagnetization factor D or by increasing the excitation frequency, because $(dHex/dt) \sim f$ up to frequency values that make eddy currents negligible.

Such devices can be also classified in parallel type and orthogonal type fluxgate sensors depending on the excitation field is parallel or perpendicular to the sensitive axis of the sensor.

Typically these devices contain solenoid systems wiring magnetic cores consisting of a permalloy or an amorphous material.

Figure 7 shows also the ring type configuration. The closed geometry of the ring core has a lower sensitivity but better noise performance, due to the absence of open ends.

A closed core made with oval geometry (race-track fluxgate sensor) lead to a lower demagnetization factor, then to higher sensitivity and less sensitivity to perpendicular fields.

In the orthogonal fluxgate the excitation coil is absent, and the sensor is excited directly by the current flowing through the core.

Three basic types of miniature fluxgate can be distinguished (Ripka & Janosěk, 2010): plane type sensors with flat coils, PCB-based devices with solenoids made by tracks and vias and 3D type sensors with micro solenoids.

While plane type sensors are typically fabricated by standard CMOS processes, MEMS microfluxgate sensors exploit advanced microfabrication technologies to realize three-dimensional coils or three-dimensional cores.

One of the first work which proposed MEMS technology for the development of fluxgate sensors has been reported by Liakopoulos & Ahn, (1999). In this work the authors presented a micro-fluxgate magnetic sensor based on micromachined toroidal type planar coils. In this fluxgate sensor a rectangular-ring shaped magnetic core has been chosen. The operation principle is based on the second harmonic. Excitation and sensing coils as well as permalloy magnetic cores were fabricated by a UV-LIGA thick photoresist process and electroplating techniques to realize a planar three-dimensional magnetic fluxgate sensor on silicon wafers. Excellent linear response over the range of -500 mT to +500 mT, with a sensitivity of 8360 VT^{-1} and a resolution of 60 nT was achieved with this device. The total response range of the sensor is -1.3 to +1.3 mT. The power consumption is around 100 mW for an operational frequency range of 1-100 kHz.

Woytasik et al. (2006) proposed an alternative fabrication process based on copper micromoulding to realize planar microcoils and microsolenoids for MEMS based magnetic sensors on flexible substrate. Figure 8 shows main steps for solenoid fabrication process.

The main steps for solenoids fabrication consist of the realization of the bottom conductor lines and of the air bridge by copper electrodeposition overflow. The second exposure process uses a gray-tone mask to vary spatially the exposure dose deposited into the photoresist and then to modulate the remaining photoresist thickness after development. The process ends with mould removal.

This technology has been employed for the realization of a micromachined fluxgate sensor (Wu & Ahn, 2008).

The sensor, schematically shown in figure 9, consists of a 30 μm thick electroplated permalloy core, with 56 excitation turns giving a total resistance of 2 Ω and 11 sensing turns. A sensitivity of 650 VT^{-1} was achieved for a 5.5-mm-long sensor with 14 mW power consumption. The noise is 32 nT/$\sqrt{Hz}$ @1Hz , and the practical resolution is 1 μT.

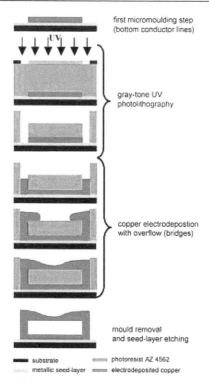

Fig. 8. Main steps of solenoid fabrication process reported by Woytasik et al. (2006).

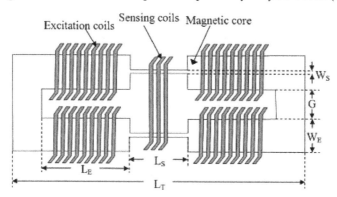

Fig. 9. Schematic view of the MEMS based fluxgate sensor proposed by Wu & Ahn, (2008).

Other examples of MEMS-based fluxgate devices have been reported by Chong Lei et al. (2009) and Kirchhoff & Büttgenbach (2010).

Most of the microfluxgate exploiting MEMS technologies are parallel sensors.

A miniature orthogonal fluxgate realized by exploiting MEMS technologies with a planar structure formed by a permalloy layer electrodeposited on a rectangular copper conductor has been reported by Zorlu et al. (2007). The sensor structure is reported in figure 10. The sensor core is only 1 mm in length and the sensor has two flat 60 turn pickup coils.

The overall dimension of the sensor chip is 1.8 × 0.8 mm, the sensitivity is 0.51 mVmT^{-1} in a linear operating range of ±200 μT. The noise was 95nT/√Hz@1Hz with an excitation power consumption of 8.1 mW.

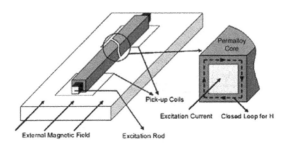

Fig. 10. Sensor structure the orthogonal microfluxgate reported by Zorlu at al. (2007).

4. Hall sensors

Hall sensors exploits Hall effect as trasduction principle to detect magnetic field. They are commonly fabricated by standard Complementary Metal-Oxide Semiconductor (CMOS) technology. In general, they are applicable in a range from 1 μT to 1 T and have a die size less than one millimeter.

These sensors can measure either constant or varying magnetic field. The frequency limit is around 1 MHz and operate well in a wide temperature range (Ripka & Tipek, 2007).

The Hall effect is based on the Lorentz force felt by charge carriers moving in a magnetic field. Figure 11 shows the schematic of the classical configuration where a thin slab of a conductor is placed in a magnetic field B_z. When a current flows in the x direction, the Lorenz force acts in the y direction, determining a charge distribution that counterbalances the force. Therefore a (Hall) voltage V_H builds up, as shown in figure 11.

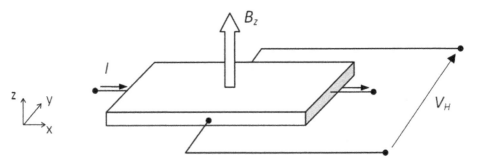

Fig. 11. Hall effect principle.

In the case of a constant current drive, the Hall voltage is given by:

$$V_H = V_{OFF} + \frac{B_z I}{e n_{2D}} \tag{7}$$

where V_{OFF} is the Hall voltage at zero magnetic field (offset voltage) and n_{2D} is the sheet charge concentration given by the product of the bulk charge concentration n and the thickness of the slab t:

$$n_{2D} = nt \tag{8}$$

If the current is constant, lower charge carriers concentration involves higher carriers speed resulting in an higher Lorentz force. Therefore low charge carrier concentration leads to high Hall voltage, thus justifying the extensive employment of semiconductor materials for Hall magnetic sensors with respect to metals.

The current related sensitivity is a key figure of merit of a Hall sensor and can be defined as:

$$S_I = \left| \frac{1}{I} \frac{\partial V_H}{\partial B_z} \right| \tag{9}$$

As the Hall voltage is related only to the z-axis magnetic field component, Hall magnetic sensors are basically uniaxial devices. Silicon based Hall sensors are widely employed, due to the suitability of integration with electronics. However, magnetic sensors based on silicon may have intrinsic limits to their sensitivity and resolution, which may limit future performance gains. In addition, they need temperature compensation circuits that can include temperature sensor and operational amplifiers (op-amps).

MEMS technology has been employed in this class of devices in order to solve some of their limitations and to find alternatives to silicon as structural material. For example polymer-based devices are interesting alternatives to silicon, particularly when the polymer materials can be functionalized for enhanced specific material properties (e.g, optical, electrical, and mechanical). Mouaziz et al. (2006) proposed the realization of SU-8 cantilevers with an integrated Hall-probe for advanced scanning probe sensing applications. To this purpose an innovative release method of polymer cantilevers with embedded integrated metal electrodes has been employed. Figure 12 shows the device fabrication process. On the silicon wafer with 0.5 µm of thermal oxide a 2 µm-thick polysilicon sacrificial layer is deposited (figure 12a). The electrodes of the Hall probe (figure 12b) and the metallic thin film electrical connections are obtained by lithographic patterning, metal deposition and liftoff (figure 12c). The device structure is obtained by lithographic patterning of two layers of SU-8 polymer: a 10µm-thick photo-structured layer for the cantilever, and a 200µm-thick layer for the chip body (figure 12d and figure 12e). The releasing method is based on dry etching of a 2µm-thick sacrificial polysilicon layer (figure 12 f).A device sensitivity of 0.05 V/AT was achieved together with a minimum detectable magnetic flux density of 9 µT/√Hz at frequencies above 1 kHz at room temperature.

Sunier et al. (2004) reported on a vertical Hall sensor with precisely defined active area fabricated by a process combining deep-RIE silicon trench etching and anisotropic TMAH silicon wet etching. The fabrication process is showed in figure 13. The main process steps are the thermal oxidation of the p-type silicon wafer (figure 13a), the definition of the trench by dry etching, the sidewall implantation and oxidation (figure 13b), another dry etching process to deepen the trenches (figure 13c), TMAH wet etching to release the bottom of the active area (figure 13d), trench oxidation, polysilicon refill and blanket etch, contact formation and metallization (figure 13e) and finally XeF$_2$ etching of the polysilicon in the tranches (figure 13f).

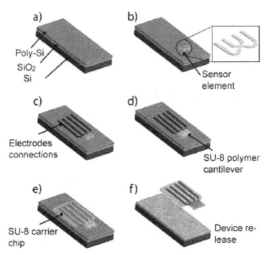

Fig. 12. Schematic illustration of the process for SU-8 cantilever with integrated electrodes reported by Mouaziz et al. (2006).

Beside a very high current related sensitivity of up to 1000 V/AT, the improved insulation from the substrate resulted in a more efficient offset compensation and then in a reduced residual offset.

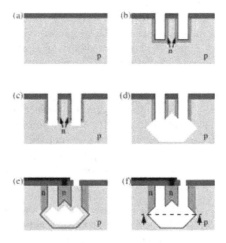

Fig. 13. Simplified fabrication process for trench Hall devices reported by Sunier et al. (2004).

Estrada (2011) proposed a three axis Hall sensor based on MEMS micromachining of SOI-wafers. Three Hall sensors embedded in a flexible polyimide carrier was obtained so that appropriate folding of the structure resulted in three Hall-probes positioned to form an orthogonally-oriented array on three faces of a millimeter-sized cube. The key fabrication process steps are as follow: (a) SOI-wafer; (b) micromachining of the active layer using either wet (TMAH) or dry (DRIE) etching; (e) patterning the Al-film needed for ohmic contacts; (f) deposition and curing of the polyimide film; g) etching of the handle wafer to reach the buried oxide film; (h) final metallization of the contact pads for possible soldering or wire bonding (figure 14).

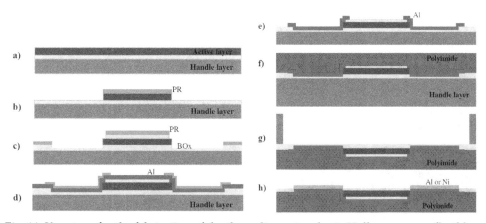

Fig. 14. Key steps for the fabrication of the three-dimensional axis Hall sensor on a flexible substrate (polyimide film) proposed by Estrada (2011).

This 3D-sensor configuration allows vector magnetic field measurements where the advantage is that all its elements have the same magnetic sensitivity of about 100 V/AT.

An integrated three axis Hall sensor based on III-V technology was fabricated by employing a micromachining technique for realizing self positioned structures (Todaro et al., 2010). The MEMS technique was applied to a GaAs-based heterostructure containing a sensing layer and a strained layer. The selective removal of a third sacrificial layer allows for the relaxation of the strained layer and the self positioning of the sensing part, which has been already processed to realize a Hall sensor element (figure 15).

The main fabrication steps are as follows: a) Epitaxial growth of a multilayer with sacrificial layer, strained layer, and sensor multilayer; (b) Photolithography and wet etching to define the mesa active region; (c) Photolithography and wet etching to define the hinge region; (d) metallization by lift off (GeAu/Ni/Au); (e) Photolithography and chemical etching to expose the edge of the sacrificial layer; (f) selective etching of the sacrificial layer and self-positioning of the structure.

Current related sensitivity of more than 1000 V/AT both for in-plane and for out-of-plane Hall sensors, demonstrates the effectiveness of this method for realizing fully integrated miniaturized high sensitivity three axis magnetic sensors.

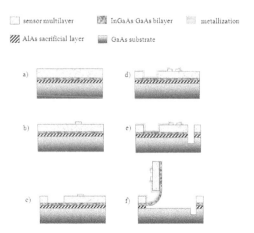

Fig. 15. Three-axis Hall sensor fabrication steps.

Figure 16 shows images of the realized three dimensional magnetic Hall sensor highlighting the out-of-plane sensor.

Fig. 16. Fully integrated three-axis Hall sensor reported by Todaro et al. (2010). a) a schematic of the three-axis Hall sensor; b) Image of a fabricated three axis Hall sensor; c) highlight of an out-of-plane Hall sensor acquired by scanning electron microscopy.

5. Applications

Magnetic field sensors based on MEMS technology have potential advantages with respect to conventional magnetic field sensors such as small size, light weight, compactness, lower power consumption. Additionally the MEMS technology achieves low-cost sensors by means of batch fabrication techniques and their potential integration with integrated circuits (IC) on a same substrate.

Conventional magnetic sensor classes presented in this paper have different application fields depending on their sensitivity range and minimum detectable magnetic field.

Resonant sensors have a magnetic range up to 1 T with a maximum resolution of 1nT, fluxgate sensors range spans from 100 pT to 1 mT, while Hall sensors have a sensitivity ranging from 1 μT to 1 T. Resonant sensors have lower resolution compared to fluxgate, however they present a wide sensitivity range and they could compete with fluxgate sensors into numerous applications for measuring magnetic fields.

Among the sensors presented in this paper, Hall sensors are the less sensitive devices. Their robustness and simple fabrication process justify their use in hundreds of applications.

Improvements in the microfabrication technologies combined with the employment of new and more performing materials as well as novel design solutions for devices on the microscale could enhance further the resolution, making them suitable for applications requiring very high sensitivity, such as in biomedical field and for the realization of new class of hand-held equipments. On the other hand this technology could help in new solutions for devices in applications requiring low sensitivity.

Magnetic field sensors on the microscale with moderate sensitivity, could be used for vehicle detection and recognition (Herrera-May et al. 2009). In fact vehicles moving over ground can generate a succession of impacts on the earth's magnetic field, that can be detected by means of magnetic perturbation using a magnetic sensor, and automatically recognize them by advanced signal processing and recognition method. In this context such sensors could be used for the measure of the speed and size of vehicles for traffic surveillance. Additionally magnetic field sensors can be used in systems containing accelerometers, gyroscopes and pressure devices for vehicle control applications (Niarchos, 2003). For example they can be employed in electronic stability program (ESP) systems to help vehicles to be dynamically stable in critical situations like hard braking and slippery surfaces.

Such microsensors can be employed in electronic compasses for sensing earth's magnetic field for GPS systems in order to provide more precise and instantaneous headings to aid navigation for air, ground and underwater systems. Additionally such devices can be used for global positioning systems (GPS) in cell phones due to the requirements of reduced size, low cost and low power consumption.

These magnetic field sensors find employment for the detection of compact ferrous objects (McFee et al. 1990). Such objects are of major concern in a number of applications. In environment science there is the need for portable sensors for mineral prospecting like measurements of magnetic properties of rocks, as well as detection of pipeline corrosion where geological ore inclusions generate typical peak magnetic induction in the range 1-1000 nT. In the military field such devices can be used in systems for the detection and mapping of hidden or unexploded ordnance (mines, bombs, and artillery shells which have a peak magnetic induction in 10-1000 nT range) as well as for detection of armored vehicles (10000 nT) or submarines (1-10 nT). The performance of these systems can be enhanced by using two or three dimensional array of sensors. This could give additional informations on the size and the depth of the buried objects.

Another application of these sensors is in non-distructive testing for a variety of evaluations including medical implants and aircraft structures, the detection of cracks and corrosion in metals. Archeology is another field requiring systems including magnetic field sensors to resolve non-invasively details, the wide range of artifacts (1-1000 nT magnetic induction range) and cultural objects. This field requires also new means of mapping prehistoric and historic sites in three dimensions rather than traditional two-dimensional methods.

High sensitivity and high resolution magnetic sensors are needed in systems for medical diagnostics. Microfluxgate sensors based on MEMS technology can be employed to build cheap and portable systems for locating metallic foreign objects in the human body (Jing et al. 2009).

Ripka, (2004) showed that fluxgate sensors can be used for mapping the distribution of ferromagnetic particles in the lungs after they are magnetized by strong DC field. Medical applications requiring precise miniaturized magnetic sensors include tracking devices and systems for monitoring magnetic markers such as magnetic "biscuits" and microbeads. Magnetic biscuits can be used for functional tests of digestive tract, while microbeads are used as markers in biotechnology. New types of fluxgate microsensors are being developed for these applications (Vopalensky et al., 2003). Also Hall magnetic sensors have been employed to visualize a magnetically market diagnostic capsule in real time inside human body (Mahfuzul-Aziz, 2008). Tracking devices using fluxgate sensors can be used for monitoring the 3-D position and also orientation of a small permanent magnet which can be attached to body or medical instrument (such as catheter). Another configuration is being used for tracking the motion of the body at further distances: signals from sensors attached to the body are collected and processed.

Typically Hall magnetic devices, due to their low sensitivity are employed for position sensing, current sensing, speed detection, electronic compasses. (Lenz & Edelstein, 2006). Silicon-based Hall sensors are widely employed, due to the suitability of integration with electronics (Popovic, 1997). However, higher sensitivity sensors can be obtained with III-V technology, allowing for applications such as biomolecular function detection (Manandhar et al., 2009). Also recently, the Scanning Hall probe microscopy (SHPM) has been developed based on III-V Hall sensors, allowing for quantitative mapping of nanoscale superconducting and ferromagnetic materials (Bending et al., 2009).

Others nowadays applications such as geomagnetic measurements, environmental disturbance measurements as well as navigation systems demand for Hall miniaturized devices capable of measuring the vector magnetic field. Beside the cumbersome solution of mounting three Hall sensors with their sensitive axis orthogonal to each other, integrated three axis devices have been developed by employing silicon technology and the so called vertical Hall effect (Schott & Popovic, 1999). However these are often characterized by either different sensitivity for each component of the magnetic field (Schott et al., 2000), or by a cross-sensitivity among the direction-components (Popovic, 1999). Furthermore, offset compensation of vertical Hall element is more difficult than in the case of lateral (planar) Hall elements. In this context new materials and device configurations could open the way to realize reliable vector magnetic field sensors to be applied in different fields.

6. Conclusion

In this paper the authors described current research status in magnetic field sensors focusing on devices fabricated by exploiting MEMS technologies. The paper presents advances in some classes of devices such as resonant sensors, fluxgate sensors and Hall sensors that take advantages from these technologies. The authors focused on the description of such microsensors including operation principle, example of realized devices, highlighting the involved fabrication technologies. Possible applications of this new class of compact devices has also been reported.

7. Acknowledgment

This work was partially supported by FIRB - Hub di ricerca italo-giapponese sulle nanotecnologie.

8. References

Bahreyni, B. (2008). Fabrication and Design of Resonant Microdevices; William Andrew, Norwich, NY, USA.

Beeby, S.; Ensell, G.; Kraft, M.; White, N. (2004). MEMS Mechanical Sensors, Artech House, Norwood, MA, USA.

Bending, S.J. & Khotkevych, V.V. (2009). Scanning Hall Probe Imaging of Nanoscale Magnetic Structures. Sensor Letters, Vol. 7, No. 3, (June 2009), pp. 503-506.

Beroulle, V.; Bertrand, Y.; Latorre, L. & Nouet, P. (2003). Monolithic piezoresistive CMOS magnetic field sensors, Sens. Actuators A, Vol. 103, No. 1-2, (January 2003), pp. 23-32, ISSN 0924-4247.

Bozorth, R.M. & Chapin, D.M. (1942). Demagnetising factors of rods, J. Appl. Phys., Vol. 13, pp. 320.

Brugger, S. & Paul, O. (2009). Field-Concentrator-Based Resonant Magnetic Sensor With Integrated Planar Coils, Journal of Microelectromechanical Systems,, Vol. 18, No.6, DECEMBER (2009), pp. 1432-1443, ISSN 1057-7157.

Chang, H.; Xue, L.; Qin, W.; Yuan, G.; Yuan, W. (2008). An integrated MEMS gyroscope array higher accuracy output, Sensors 2008, Vol. 8, pp. 2886-2899.

Elwenspoek, M. & Wiegerink, R. (2001). Mechanical Microsensors, Springer-Verlag: Berling, Heidelberg, Germany.

Emmerich, H. & Schöfthaler, M. (2000). Magnetic field measurements with a novel surface micromachined magnetic-field sensor, IEEE Trans. Electron Dev., Vol. 47, No. 5, (May 2000), pp. 972-977, ISSN 0018-9383.

Estrada, H.V. (2011). A MEMS-SOI 3D-magnetic field sensor, 24th International Conference on Micro Electro Mechanical Systems (MEMS). IEEE 2011, pp. 664-667.

Herrera-May, A. L.; García-Ramírez, P.J.; Aguilera-Cortés, L.A.; Martínez-Castillo, J.M.; Sauceda-Carvajal A.; García-González, L. & Figueras-Costa, E. (2009). A resonant magnetic field microsensor with high quality factor at atmospheric pressure, J. Micromech. Microeng. , Vol. 19, No. 1, (January 2009),pp. 015016 -015026, ISSN 1057-7157.

Herrera-May, A. L.; Aguilera-Cortés, L.A.; García-Ramírez, P.J.; Manjarrez, E. (2009). Resonant magnetic field sensors based on MEMS technology, Sensors 2009, Vol. 9, pp. 7785-7813, ISSN 1424-8220.

Herrera-May, A. L.; Aguilera-Cortés, L.A.; García-Ramírez, P.J.; Mota-Carrillo, N.B.; Padrón-Hernández, W.Y. & Figueras, E. (2010). Development of resonant magnetic field microsensors: challenges and future applications, Microsensors, Intech publisher, ISBN 978-953-307-170-1.

Hull, R. Properties of Crystalline Silicon; Institution of Electrical Engineers: London, UK, 1999.

Jing, D.; Luo, E.; Shen, G.; Cai, J.; Tang, C.; Yan, Y.; Jing, B. (2009). Fast Method of Locating Metallic Foreign Body in the Human Body. The Ninth International Conference on Electronic Measurement & Instruments, ICEMI, pp. 4-843.

Kádár, Z.; Bossche, A.; Sarro, P.M. & Mollinger, J.R. (1998). Magnetic-field measurements using an integrated resonant magnetic-field sensor, Sens. Actuators A, Vol. 70 No. 3, (October 1998),, pp. 225-232, ISSN 0924-4247.

Keplinger, F.; Kvasnica, S.; Jachimowicz, A.; Kohl, F.; Steurer, J. & Hauser, H. (2004). Lorentz force based magnetic field sensor with optical readout, Sens. Actuators A, Vol. 110, No. 1-3, (February 2004), pp. 112-118, ISSN 0924-4247.

Kirchhoff, M.R. & Büttgenbach, S. (2010). MEMS fluxgate magnetometer for parallel robot application, Microsyst Technol , Vol. 16, pp. 787–790.

Kyynäräinen, J.; Saarilahti, J.; Kattelus, H.; Kärkkäinen, A.; Meinander, T.; Oja, A.; Pekko, P.; Seppä, H.; Suhonen, M.; Kuisma, H.; Ruotsalainen, S. & Tilli, M. (2008). A 3D micromechanical compass, Sensors and Actuators A, Vol. 142, pp. 561–568.

Lei, C.; Wang, R.; Zhou, Y. & Zhou, Z. (2009). MEMS micro fluxgate sensors with mutual vertical excitation coils and detection coils, Microsyst Technol., Vol. 15, pp. 969–972.

Lenz, J. & Edelstein, A.S. (2006). Magnetic sensors and their applications, IEEE Sensors Journal, Vol. 6, pp. 631-649.

Li, Y.; Zheng, Q.; Hu, Y. & Xu, Y. (2011). Micromachined Piezoresistive Accelerometer Based on a Asymmetrically Gapped Cantilever. Journal of Microelectromechanical Systems, Vol. 20, No. 1, (February 2011), pp. 83-94, ISSN 1057-7157.

Liakopoulos, T.M. & Ahn, C.H. (1999). A micro-fluxgate magnetic sensor using micromachined planar solenoid coils, Sensors and Actuators, Vol. 77, pp. 66–72.

Mahfuzul-Aziz, S.; Grcic, M. & Vaithianathan, T. (2008). A real-time tracking systems for an endoscopic capsule using multiple magnetic sensors. In Smart Sensors and Sensing Technology; Mukhupadhyay, S.C., Gupta, G.S., Eds.;. Springer-Verlag: Heidelberg, Germany, pp. 201-218.

Manandhar, P.; Chen, K.-S.; Aledealat, K.; Mihajlović, G.; Yun, S.; Field, M.; Sullivan, G.J.; Strouse, G.F.; Chase, P.B.; von Molnár, S. & Xiong, P. (2009). The detection of specific biomolecular interaction with micro-Hall magnetic sensors, Nanotechnology, Vol. 20, pp. 355501.

McFee J. E., Das Y., Ellingson R.O., (1990) Locating and Identifying Compact Ferrous Objects, IEEE Trans. Geosci. Remote Sens., Vol. 28, pag 182.

Mian, A. & Law, J. (2010). Geometric Optimization of a Van Der Pauw Structure Based MEMS Pressure Sensor. Microsystem Technologies, Vol. 16, No. 11, (November 2010), pp. 1921-1929, ISSN 0946-7076.

Mouaziz, S.; Boero, G; Popovic, R.S.; Brugger, J. (2006) Polymer-Based Cantilevers with Integrated Electrodes. Journ. Microelectromechanical Systems, Vol. 15, pp. 890

Niarchos, D. (2003). Magnetic MEMS: key issues and some applications. Sens. Actuat. A 2003, Vol. 109, pp. 166-173.

Popovic, R.S. (1997). Hall Devices for Magnetic Sensor Microsystems. IEEE Transducers'97, pp. 377.

Popovic, R.S. (1999). Novel Hall Magnetic Sensors and their Applications. EUROSENSORS XIII, 13th European Conference on Solid-State Transducers, The Hague, The Netherlands, pp. 1041-44.

Primdahl, F. (1979). The Fluxgate magnetometer, J. Phys. E: Sci. Instrum. Vol. 12.

Ren, D.; Wu, L.; Yan, M.; Cui, M.; You, Z. & Hu, M. (2009). Design and Analyses of a MEMS Based Resonant Magnetometer, Sensors 2009, Vol.9, pp. 6951-6966.

Ripka, P. (2001). Magnetic Sensors and Magnetometers, Artech House, ISBN 1-58053-057-5 Boston, USA

Ripka, P. (2004). Biomedical Application of Fluxgate Sensors. Progress in Electromagnetic Research Symposium 2004, March 28-31.

Ripka, P. & Tipek, A. (2007). Modern sensors Handbook, Wiley.

Ripka, P. & Janosěk, M. (2010). Advances in magnetic field sensors, IEEE Sensors Journal, Vol. 10, no. 6, (June 2010), pp. 1108-1116.

Sabaté, N.; Vogel, D.; Gollhardt, A.; Keller, J.; Cané, C.; Gràcia, I.; Morante, J.R.; Michel, B. (2007). Residual stress measurement on a MEMS structure with high-spatial resolution. J. Microelectromech. Syst., Vol.6, pp. 365-372.

Schott, Ch. & Popovic R.S. (1999). Integrated 3D Hall magnetic field sensor, Transducers'99 Proc. Conf. on Solid-State Sensors and Actuators, pp. 169-71.

Schott, C.; Besse, P.A.; Popovic, R.S. (2000). Planar Hall Effect in the Vertical Hall Sensor. Sensors and Actuators, Vol. 85, pp. 111-115.

Singh, J.; Teo, J.H.S.; Xu, Y.; Premachandran, C.S.; Chen, N.; Kotlanka, R.; Olivo, M.; Sheppard, C.J.R. (2008). A two axes scanning SOI MEMS micromirror for endoscopic bioimaging. J. Micromech. Microeng. 2008, Vol. 18, pp. 025001.

Sunier, R.; Monajemi, P.; Ayazi, F.; Vancura, T.; Baltes, H.; Brand, H. (2004). Precise release and insulation technology for vertical Hall sensors and trench-defined MEMS. Sensors, Proceedings of IEEE (2004), pp 1442.

Sunier, R.; Vancura, T.; Li, Y.; Kay-Uwe, K. & Baltes, H.; Brand, O. (2006). Resonant magnetic field sensor with frequency output. J. Microelectromech. Syst., Vol. 15, pp. 1098-1107.

Tapia, J.A.; Herrera-May, A.L.; García-Ramírez P.J.; Martinez-Castillo, J.; Figueras, E.; Flores, A. & Manjarrez, E. (2011). Sensing magnetic flux density of artificial neurons with a MEMS device, Biomed Microdevices, Vol. 13, pp. 303–313.

Todaro, M.T.; Sileo, L.; Epifani, G.; Tasco, V.; Cingolani, R.; De Vittorio, M.; Passaseo, A. (2010). A Fully integrated GaAs-based three-axis Hall magnetic sensor exploiting self-posizioned strain released structures. J. Micromech. Microeng., Vol 20, pp. 105013.

Tucker, J.; Wesoleck, D. & Wickenden, D. (2000). An integrated CMOS MEMS xylophone magnetometer with capacitive sense electronics. In 2000 NanoTech, Houston, Texas, USA, 9-12

Vopalensky, M.; Ripka, P.; Platil, A. (2003). Precise Magnetic Sensors. Sensors & Actuators A, Vol 106, pp. 38-42.

Weaver, W. Jr.; Timoshenko, S.P.; Young, D.H. (1990). Vibration Problems in Engineering, 5th ed., Wiley: New York, NY, USA.

Wickenden, D.K.; Champion, J.L.; Osiander, R.; Givens, R.B.; Lamb, J.L.; Miragliotta, J.A.; Oursler, D.A. & Kistenmacher, T.J. (2003). Micromachined polysilicon resonating xylophone bar magnetometer, Acta Astronautica, Vol. 52, pp. 421-425.

Woytasik, M.;Grandchamp, J.-P.; Dufour-Gergam, E.; Gilles, J.-P.; Megherbi, S.; Martincic, E.; Mathias, H. & Crozat, P. (2006). Two- and three-dimensional microcoil fabrication process for three-axis magnetic sensors on flexible substrates, Sens. Act. A: Phys., Vol. 132, pp. 2-7.

Wu, P.-M. & Ahn, C.H. (2008). Design of a low-power micromachined fluxgate sensor using localized core saturation method, IEEE Sensors J., Vol. 8, pp. 308–313.

Zorlu, O.; Kejik, P. & Popovic, S. (2007). An orthogonal fluxgate-type magnetic microsensor with electroplated Permalloy core, Sens. Act. A, Phys., Vol. 135, pp. 43–49.

Permissions

The contributors of this book come from diverse backgrounds, making this book a truly international effort. This book will bring forth new frontiers with its revolutionizing research information and detailed analysis of the nascent developments around the world.

We would like to thank Dr. Kevin Kuang, for lending his expertise to make the book truly unique. He has played a crucial role in the development of this book. Without his invaluable contribution this book wouldn't have been possible. He has made vital efforts to compile up to date information on the varied aspects of this subject to make this book a valuable addition to the collection of many professionals and students.

This book was conceptualized with the vision of imparting up-to-date information and advanced data in this field. To ensure the same, a matchless editorial board was set up. Every individual on the board went through rigorous rounds of assessment to prove their worth. After which they invested a large part of their time researching and compiling the most relevant data for our readers. Conferences and sessions were held from time to time between the editorial board and the contributing authors to present the data in the most comprehensible form. The editorial team has worked tirelessly to provide valuable and valid information to help people across the globe.

Every chapter published in this book has been scrutinized by our experts. Their significance has been extensively debated. The topics covered herein carry significant findings which will fuel the growth of the discipline. They may even be implemented as practical applications or may be referred to as a beginning point for another development. Chapters in this book were first published by InTech; hereby published with permission under the Creative Commons Attribution License or equivalent.

The editorial board has been involved in producing this book since its inception. They have spent rigorous hours researching and exploring the diverse topics which have resulted in the successful publishing of this book. They have passed on their knowledge of decades through this book. To expedite this challenging task, the publisher supported the team at every step. A small team of assistant editors was also appointed to further simplify the editing procedure and attain best results for the readers.

Our editorial team has been hand-picked from every corner of the world. Their multi-ethnicity adds dynamic inputs to the discussions which result in innovative outcomes. These outcomes are then further discussed with the researchers and contributors who give their valuable feedback and opinion regarding the same. The feedback is then collaborated with the researches and they are edited in a comprehensive manner to aid the understanding of the subject.

Apart from the editorial board, the designing team has also invested a significant amount of their time in understanding the subject and creating the most relevant covers. They scrutinized every image to scout for the most suitable representation of the subject and create an appropriate cover for the book.

The publishing team has been involved in this book since its early stages. They were actively engaged in every process, be it collecting the data, connecting with the contributors or procuring relevant information. The team has been an ardent support to the editorial, designing and production team. Their endless efforts to recruit the best for this project, has resulted in the accomplishment of this book. They are a veteran in the field of academics and their pool of knowledge is as vast as their experience in printing. Their expertise and guidance has proved useful at every step. Their uncompromising quality standards have made this book an exceptional effort. Their encouragement from time to time has been an inspiration for everyone.

The publisher and the editorial board hope that this book will prove to be a valuable piece of knowledge for researchers, students, practitioners and scholars across the globe.

List of Contributors

Mattia Butta
Kyushu University, Japan

Renu Choithrani
Department of Physics, Barkatullah University, Bhopal, India

Christophe Coillot and Paul Leroy
LPP Laboratory of Plasma Physics, France

Takaya Inamori and Shinichi Nakasuka
The University of Tokyo, Japan

Chengliang Huang and Xiao-Ping Zhang
Ryerson University, Canada

Guillermo Rivero, Marta Multigner and Jorge Spottorno
Instituto de Magnetismo Aplicado (Universidad Complutense de Madrid), Spain

Massimo De Vittorio
National Nanotechnology Laboratory (NNL), Istituto Nanoscienze-CNR, Italy
Center for Biomolecular Nanotechnologies UNILE, Istituto Italiano di Tecnologia, Italy
University of Salento, Lecce, Italy

Maria Teresa Todaro
National Nanotechnology Laboratory (NNL), Istituto Nanoscienze-CNR, Italy

Leonardo Sileo
National Nanotechnology Laboratory (NNL), Istituto Nanoscienze-CNR, Italy
Center for Biomolecular Nanotechnologies UNILE, Istituto Italiano di Tecnologia, Italy

Printed in the USA
CPSIA information can be obtained
at www.ICGtesting.com
JSHW011339221024
72173JS00003B/175

9 781632 383099